3rd International Conference on Materials and Intelligent Manufacturing (ICMIM 2019)

AA007751

IOP Conference Series: Materials Science and Engineering
Volume 654

Incheon, South Korea
19 - 22 August 2019

ISBN: 978-1-5108-9849-1
ISSN: 1757-8981

Printed from e-media with permission by:

Curran Associates, Inc.
57 Morehouse Lane
Red Hook, NY 12571

proceedings.com

Some format issues inherent in the e-media version may also appear in this print version.

This work is licensed under a Creative Commons Attribution 3.0 International Licence.
Licence details: http://creativecommons.org/licenses/by/3.0/.

No changes have been made to the content of these proceedings. There may be changes to pagination and minor adjustments for aesthetics.

Printed with permission by Curran Associates, Inc. (2020)

For permission requests, please contact the Institute of Physics at the address below.

Institute of Physics
Dirac House, Temple Back
Bristol BS1 6BE UK

Phone: 44 1 17 929 7481
Fax: 44 1 17 920 0979

techtracking@iop.org

Additional copies of this publication are available from:

Curran Associates, Inc.
57 Morehouse Lane
Red Hook, NY 12571 USA
Phone: 845-758-0400
Fax: 845-758-2633
Email: curran@proceedings.com
Web: www.proceedings.com

Table of contents

Volume 654

3rd International Conference on Materials and Intelligent Manufacturing (ICMIM)

19–22 August 2019, Incheon, South Korea

Accepted papers received: 19 September 2019
Published online: 28 October 2019

Preface

Papers

Mechanical Engineering

Nanometer Materials

Nanocomposites

Material Processing

Material Science

IOP Conf. Series: Materials Science and Engineering **654** (2019) 011001 doi:10.1088/1757-899X/654/1/011001

Preface

2019 3rd International Conference on Materials and Intelligent Manufacturing (ICMIM 2019) was held in Inha University, Incheon, South Korea during August 19-22, 2019. With its workshop 2019 2nd International Conference on Nano Science and Technology (ICNST 2019) held together, the conferences were great success. The conferences provide a wide platform to the conference participants to share their research achievements and communicate with each other. And the distinguished keynote speakers of the conference presented their excellent academic research at the conference which has promoted the academic atmosphere.

On behalf of the organizing committee, we are pleased to present you with this volume. This set of conference proceedings presents a selection of papers submitted to the conference from various universities, research institutes and industries. The topics include Mechanical Engineering, Nanometer Materials, Nanocomposites, Material Processing, Material Science and so on. The reviewing process was a challenge process that relies on goodwill of those people involved in this field. We invited more than 50 researchers from related fields to review papers for the publication in the IOP conference proceedings.

Meanwhile, we would like to thank all the author submitted papers to the conferences and thank you all the reviewers for their time and effort in reviewing articles. Finally, it is a great honor to thank you all of behind staff who made their constant effort to make sure the conference process can go smoothly and the proceeding team who have dedicated their constant support and countless time to bring these scratches into this volume.

Prof. Han-Yong Jeon
Inha University, South Korea

Content from this work may be used under the terms of the Creative Commons Attribution 3.0 licence. Any further distribution of this work must maintain attribution to the author(s) and the title of the work, journal citation and DOI.
Published under licence by IOP Publishing Ltd

IOP Conf. Series: Materials Science and Engineering **654** (2019) 011001 doi:10.1088/1757-899X/654/1/011001

STATEMENT OF PEER REVIEW

All papers published in this volume of *IOP Conference Series: Materials Science and Engineering* have been peer reviewed through processes administered by the proceedings Editors. Reviews were conducted by expert referees to the professional and scientific standards expected of a proceedings journal published by IOP Publishing.

IOP Conf. Series: Materials Science and Engineering **654** (2019) 011002 doi:10.1088/1757-899X/654/1/011002

Conference General Co-Chairs

Prof. Han-Yong Jeon, Inha University, South Korea
Prof. Vasudevanpillai Biju, Hokkaido University, Japan
Prof. Jun Ding, National University of Singapore, Singapore

Program Co-Chairs

Prof. Yoshihiro Sugi, Gifu University, Japan
Prof. Shen-Ming Chen, National Taipei University of Technology, Taiwan
Assoc. Prof. Pan Jisheng, National University of Singapore, Singapore

Technical Committees

Prof. Xu Guang, Wuhan University of Science and Technology, China
Prof. Manoj Kumar Singh, CSIR-Central Institute of Mining & Fuel Research, India
Prof. Xu Guang, Wuhan University of Science and Technology, China
Prof. N. Ethiraj, Dr. M.G.R Educational and Research Institute-University, India
Prof. Konovalov Sergey,Samara National Research University
Prof. Shibin Li, University of Electronic Science and Technology, China
Prof. Liang Hao, Tianjin University of Science and Technology, China
Prof. Aidy Ali, National Defence University of Malaysia, Malaysia
Prof. Wengang Bi, Hebei University of Technology, China
Prof. Nacer Badi, University of Tabuk, Saudi Arabia
Prof. Osman ADIGUZEL, Firat University, Turkey
Assoc. Prof Kitsakorn Locharoenrat， King Mongkut's Institute of Technology Ladkrabang, Thailand
Assoc. Prof. Anika Zafiah Mohd Rus, Universiti Tun Hussein Onn Malaysia, Malaysia
Assoc. Prof. Muthuramalingam Thangaraj, SRM Institute of Science and Technology, India
Assoc, Prof. S. Balasivanandha Prabu, Anna University, INDIA
Prin. Dr. Vimal Joshi, Principal, Sardar Patel University, India
Dr. Mazeyar Parvinzadeh Gashti, Chief Scientist, PRE Labs Inc., Canada
Dr. Radu Godina, University of Beira Interior, Portugal
Dr. Alain Kusmoko, University of Wollongong, Australia
Dr. Salman Pervaiz, Rochester Insititute of Technology- Dubai Campus, Dubai, UAE
Dr. Oh Pei Ching, Universiti Teknologi PETRONAS Malaysia, Malaysia
Dr. Vilas Haridas Gaidhane, BITS Pilani, Dubai International Academic City, UAE
Dr. Lai Ti Gew, Sunway University, Malaysia
Dr. Mozaffar Salimi, Malek Ashtar University of Technology, Iran
Dr. Pratik Shah, Yonsei University, South Korea

Content from this work may be used under the terms of the Creative Commons Attribution 3.0 licence. Any further distribution of this work must maintain attribution to the author(s) and the title of the work, journal citation and DOI.
Published under licence by IOP Publishing Ltd

IOP Conf. Series: Materials Science and Engineering **654** (2019) 011003 doi:10.1088/1757-899X/654/1/011003

Peer review statement

All papers published in this volume of *IOP Conference Series: Materials Science and Engineering* have been peer reviewed through processes administered by the proceedings Editors. Reviews were conducted by expert referees to the professional and scientific standards expected of a proceedings journal published by IOP Publishing.

Content from this work may be used under the terms of the Creative Commons Attribution 3.0 licence. Any further distribution of this work must maintain attribution to the author(s) and the title of the work, journal citation and DOI.
Published under licence by IOP Publishing Ltd

IOP Conf. Series: Materials Science and Engineering **654** (2019) 012001 doi:10.1088/1757-899X/654/1/012001

Chatter Identification in End Milling Process Based on Cutting Force Signal Processing

M Q Tran[1,2], **M K Liu**[1,2,*]

[1]Deparment of Mechanical Engineering, National Taiwan University of Science and Technology, Taipei, Taiwan
[2]Center for Cyber-physical System Innovation, National Taiwan University of Science and Technology, Taipei, Taiwan

[*]Corresponding author: mkliu@mail.ntust.edu.tw

Abstract. During milling process, chatter is one of the major limitations of productivity, it has more negative effects on the surface finish, dimensional accuracy and tool life. It is thus necessary to detect and avoid chatter occurrence. In this study, a chatter detection approach based on cutting force signal is presented. The milling force signals were measured by cutting force sensor and then analyzed by using time-frequency analysis approaches such as short-time Fourier transform (STFT) and continuous wavelet transform (CWT) in order to identify chatter. The different sets of cutting conditions were utilized to recognize chatter presence. The spectrogram and scalogram results of the force signals represented a great difference in frequency content between stable and unstable cuttings. The analysis results were then validated by cutting force behavior in the time domain.

1. Introduction

Machine tool chatter is a self-excited vibration that causes machining instability, it results in reduced surface quality and increased tool wear during cutting process [1, 2]. Chatter detection is therefore very important to obtain stable cutting parameters such as feed rate, spindle speed, and depth of cut (DOC). A stability lobe diagram could be a useful way to predict chatter [3]. However, the stability lobe diagram would inevitably misinterpret the important features of nonlinear vibration because most cutting processes contain rich nonlinear characteristics, and the linear approximation [4]. It is therefore necessary to analyze the signal in time-frequency domain.

In order to distinguish chatter occurrence, some machining process signals have to be measured by using different techniques such as in-process real-time techniques based on force and displacement signals or off-line machined surface characterization [5]. The Fourier transform (FT) is widely used to transform time-domain signal into frequency domain signal and it is reportedly suitable for the stationary signal process, especially when the feature components of signals are obvious in magnitude [6]. For non-stationary signal, the wavelet transform as a flexible method is suitable for detection of a sudden frequency change in a signal [7]. Chatter is associated with amplitude modulation and an increase in the signal energy due to the high variation in the cutting force compared with chatter free conditions [8, 9]. There are a lot of studies using fast Fourier transform (FFT) to detect cutting chatter, this method is well known as a standard method for observing signals in the frequency domain [10]. Based on the Fourier transform analysis, the FFT is the main method that can produce an acceptable discrete spectrum analysis transferring from the time domain to the frequency domain. Schmitz et al., 2002 utilized FFT of the once per revolution sampled audio signal in milling to identify chatter [11].

Content from this work may be used under the terms of the Creative Commons Attribution 3.0 licence. Any further distribution of this work must maintain attribution to the author(s) and the title of the work, journal citation and DOI.
Published under licence by IOP Publishing Ltd

Huang P et al., 2013 used the cutting force signal to detect chatter in milling operations. Based on Fourier transform, chatter can be ascertained and the chatter frequency can be obtained [12]. For nonstationary signal, the wavelet process gives better resolution capabilities in both time and frequency domain which is advantageous in the analysis of non-stationary time series arising in metal cutting [13]. To detect cutting chatter via force sensors based on wavelet transform, the cutting force signals are decomposed into a set of approximation coefficient and detail coefficients. The former is a low-frequency component, and the latter is high-frequency components of the signal that is called multilevel signal decomposition [14]. Khraisheh et al. reported that the WT can clearly define the boundary between the transient and the steady-state. By using the wavelet transform to determine the local dominant frequencies, system parameters can be accurately estimated [15]. The decomposition signals of cutting force can be obtained using the wavelet transform, and the abrupt changing amplitude of the decomposed signals can be seen as valid evidence to assess whether chatter occurred [7].

The present work focuses on the indirect approach to detect chatter in milling. For this approach, the tool conditions are not captured directly but estimated from the measurable signal feature. The milling force signals in the direction perpendicular to the machined surface were analyzed through time-frequency analysis. By conducting experiments with different cutting conditions, the effects of spindle speed, feed rates and depth of cut on milling stability were studied. The differentiation between various cutting states was made on the basis of qualitative features of fast Fourier transform, short-time Fourier transform, and CTW based on time-frequency plots. The cutting force behavior in the time-domain was utilized to validate the analysis result.

2. Signal analysis approaches

Figure 1 represents the flowchart of the utilized signal analysis approaches used to determine the presence of chatter. These methods used for evaluating the vibration signals involved in the transformation of the cutting force signals from time to frequency domain are Fast Fourier Transform (FFT), Short-Time Fourier Transform (STFT), and Continuous Wavelet Transform (CWT). Through these methods, it was possible to view properties or characteristic information of the cutting process hidden in the time domain [6]. Fast Fourier transform. The Fourier transform (FT) transforms the signal $x(t)$ from the time domain to the frequency domain $\hat{x}(\omega)$ described in Eq. (1).

$$\hat{x}(\omega) = \int_{-\infty}^{+\infty} x(t)e^{-j\omega t}dt \tag{1}$$

Short-Time Fourier transform could overcome the limitation of FFT as mentioned previously, Gabor introduced a sliding window function to the FT and obtained a localized time-frequency [10] described in Eq. (2), where $g(t)$ is window function, $\phi_{\tau,\omega}$ is a complex function representing the phase and magnitude of the signal over time and frequency.

$$\left[x(t)\phi_{\tau,\omega}\right] = \int_{-\infty}^{+\infty} x(t)(t-\tau)e^{-j\omega t}dt \tag{2}$$

Wavelet is advantageous in the analysis of non-stationary time series arising in metal cutting due to its better resolution in both time and frequency domain, according to Burger et al. [13]. CWT of the signal $x(t)$ is defined by Eq. (3), in which u is the position parameter of wavelet; s is the function scale parameter of wavelet function, $s, u \in R$, $s > 0$; W_ψ is wavelet function, and $\psi(t)$ is scale function. The scale parameter could be converted to their frequency counterpart in the scalogram [16].

$$W_\psi x(s,u) = \frac{1}{\sqrt{s}}\int_{-\infty}^{+\infty} x(t)\psi\left(\frac{t-u}{s}\right)dt \tag{3}$$

Finally, the sampled once per revolution cutting force signal was used to verify the results of the chatter detection methods.

IOP Conf. Series: Materials Science and Engineering **654** (2019) 012001 doi:10.1088/1757-899X/654/1/012001

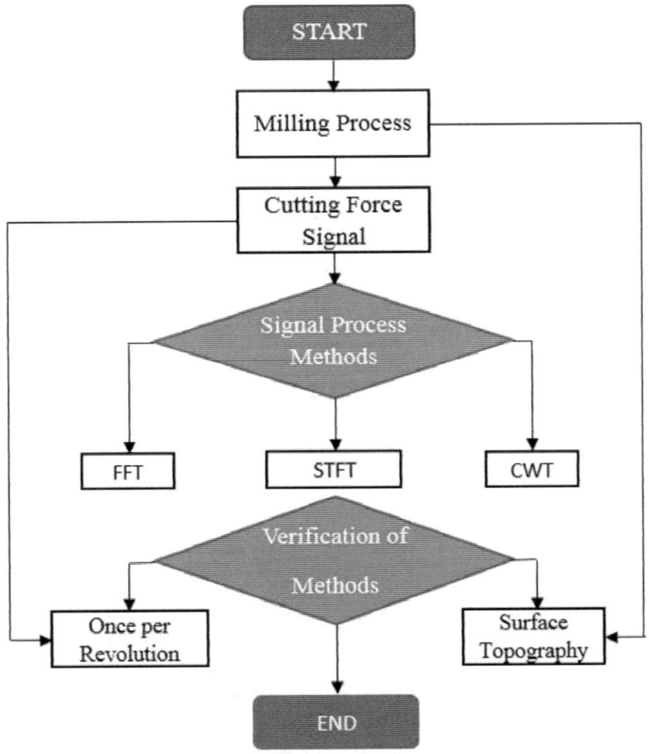

Figure 1. Flowchart of signal analysis approaches.

3. Experimental Setup

To investigate the proposed methodology for chatter identification, end-milling experiments were conducted using a 3-axis CNC milling. Al6061-T6 rectangular workpiece with dimensions 30mm x 30mm x 40mm was used to conduct the experiments. A carbide end mill cutter with a diameter of 12 mm, helix angle of 26 and two flutes was utilized. The cutting force signals were measured by a Kistler dynamometer (type 9129AA) mounted between the workpiece and workbench. The whole experimental setup is shown in Figure 2. A series of end milling experiments were conducted considering four different cutting conditions shown in Table 1. Then the cutting force signals in the feed direction were measured.

Figure 2. Experimental setup

Table 1. Cutting parameter for chatter tests.

Cutting parameters	Case 1	Case 2	Case 3	Case 4
Spindle speed (rpm)	4500	4500	2500	2500
Depth of cut (mm)	0.8	1.4	1.0	1.0
Feed rate (mm/min)	150	150	100	250

4. Results and discussions

Figure 3 shows the force spectrum plot after applying Fast Fourier Transform on the cutting force data. In the first case, Figure 3(a), shows the tooth passing frequency of 150 Hz and its harmonics. It should be noted that the tool passing and their harmonics are expected to be present in the force spectrum. There are no peaks beyond the frequency corresponding to the first or higher-order modes which are a typical chatter signature indicating that the process is stable. There are no (side lobes) disturbances near the natural frequencies of the tool i.e. 75 Hz, 150 Hz, and 225 Hz. Figure 3(b), however, shows the cutting force in the feed direction in the frequency-domain at 4500 rpm and 1.4 mm depth of cut (case 2). It is clearly noted that there are high peaks (between 4000 and 5000 Hz) beyond the tooth passing frequencies and their harmonics (natural frequencies) which signifies unstable condition thereby indicating the presences of chatter. The spectrum of the signal in case 3 is similar to case 1, in which it represents the tooth passing frequency of 83.3 Hz and its harmonics. The presence of chatter in Figure 3(d) can also be seen from the presences of disturbances near the natural frequencies.

However, it is not clear to define those frequencies whether they belong to chatter frequency. The Fourier-based methods are nonetheless based on the assumption that the analyzed signal is stationary in the time domain. The globally inclusive of information makes it difficult to provide any information about the individual frequency event of the signal. It is accordingly not suitable for many non-stationary systems when their transient responses are analyzed.

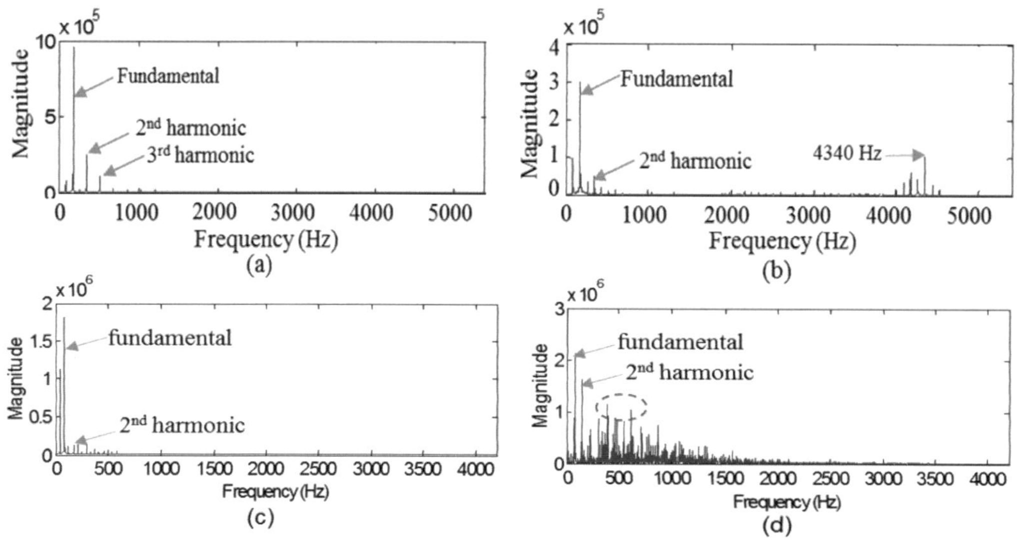

Figure 3. Fourier transform of force signals of a) case 1, b) case 2, c) case 3, and d) case 4.

In this study, the Hanning window is used in the modeling of STFT, the spectrogram results are illustrated in Figure 4. Figure 4(a) shows that the process is chatter free since the frequency domain only shows the tooth passing frequencies at 175 Hz and their fundamental frequencies. The increase of the depth of cut from 0.8 mm to 1.4 mm resulted in the occurrence of chatter as shown in Figure 4(b). This effect of depth of cut can be explained by the high energy signal at approximately 4000 Hz. When different feed rate is used, the signal is scattered when feed is 250 mm/min as shown in Figure 4(d), this clearly supports the results found above by using the FFT. For the STFT approach, even though the Hanning window was applied with the appropriate length for getting better resolution, the time resolution is still not really good in the time-domain. Furthermore, this analysis method takes a lot of time to simulate, and it is still difficult to detect unsteady changes quickly and accurately. Therefore, STFT analysis is unsuitable for real-time recognition of chatter vibrations. As mentioned, the limitation of the STFT approach can be solved by using the continuous wavelet to efficiently determine the signal frequency and exact time when chatter occurs.

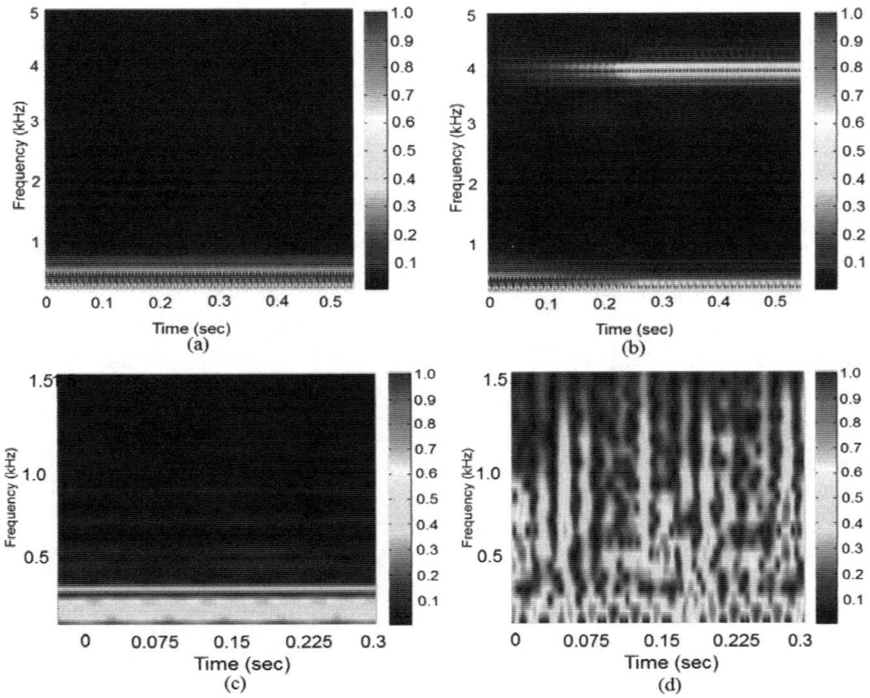

Figure 4. STFT of force signals: (a) case 1, (b) case 2, (c) case 3 and (d) case 4.

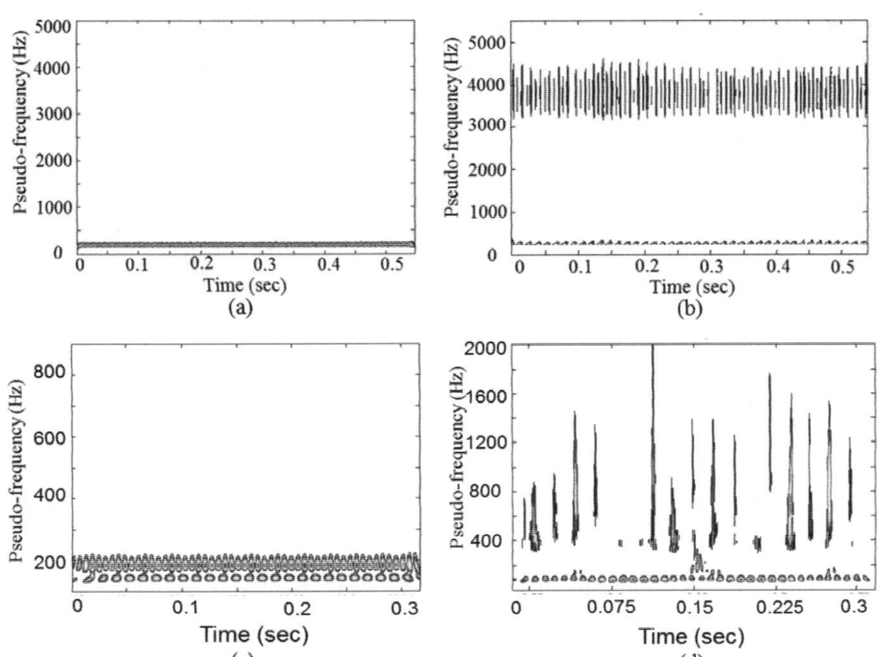

Figure 5. CWT of force signals of (a) case 1, (b) case 2, (c) case 3 and (d) case 4.

The scalogram results of the continuous wavelet transform applied on force signals for four cases are illustrated in Figure 5. The tooth passing frequency, or fundamental frequency, is clearly shown at 175 Hz in the CWT 2-D contour shown in Figure 5(a). It is observed that there is no dominant frequency so far from those frequencies. On the other hand, as the depth of cut is increased, the wavelet contour in Figure 5(b) indicates an unstable cutting behavior, which has a set of high frequencies oscillating around 4000 Hz and a tooth passing frequency of 175 Hz. The high frequencies around 4000 Hz could be defined as chatter frequencies caused by the interaction between the cutting tool and the workpiece. These high chatter frequencies are generated by the regenerative effect; that is, multiple cutting behaviors happening within one single cut due to the surface waviness on the workpiece. Even though the depth of cut has a strong influence on the generation of chatter, the results show that the change in feed rate significantly affects the signal. Using identical depth of cut of 1mm and different feed rate of 100 mm/min and 250 mm/min, as depicted in Figure 5(d) machining process instability is clearly shown by the presences of the scattered signals around the natural frequencies.

5. Verification of chatter detection method

The behavior of once per revolution sampled cutting force signal could be used to analyze the behavior of stable and unstable signals in time-domain. In this study, standard deviation values of once-per-revolution sampled cutting force data can be used to indicate a difference between stable and unstable cases. Figure 6(a) and (c) represent the once-per-revolution sampled cutting force data of case 1 and 3 respectively. Both of them show the repetitive behavior from one to the next revolution with small values of standard deviation illustrated in Table 2. Evidently, this behavior indicates maintained stable operation. On the other hand, sampled force data in Figure 6(b) and (d) show different behavior with the quasi-periodic, as a result of increasing the standard deviation values in both cases shown in Table 2. This demonstrates unstable behaviors in case 2 and 4. All the analyses in time domain using once-per-revolution sampled cutting force technique are also in good accord with the results obtained from the proposed time-frequency approach.

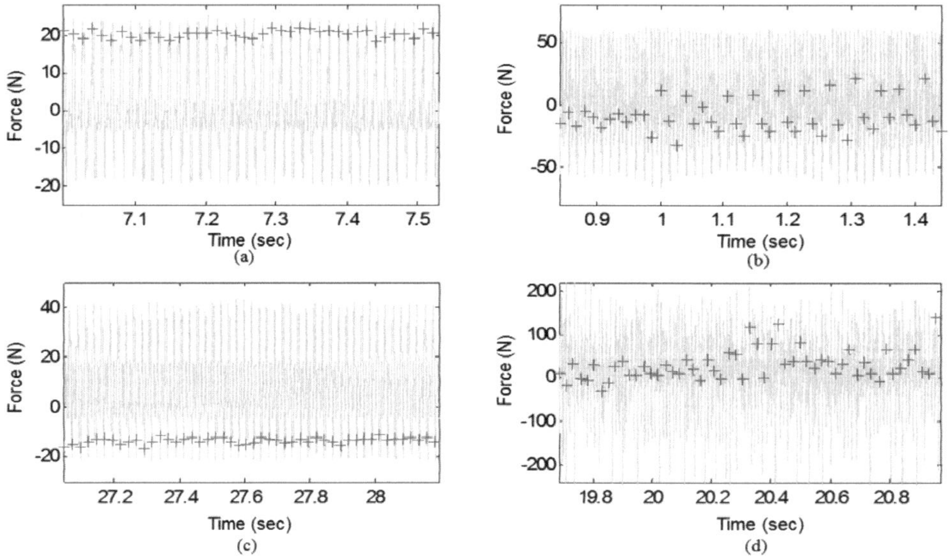

Figure 6. Sampled force data in time domain: (a) case 1, (b) case 2, (c) case 3 and (d) case 4 ("+" once-per-revolution sampled cutting force).

Table 2. Standard deviation values.

Case 1	Case 2	Case 3	Case 4
0.7824	10.8110	0.9367	25.8462

6. Conclusions

In this study, the chatter vibration was detected in-process for the milling force signals. Different approaches were utilized for signal processing such as Fourier transform, short-time Fourier transform and continuous wavelet transform to detect the chatter. It can be concluded that the frequency content of the force signals on spectrogram and scalogram are much different between stable and unstable cuttings. The results from STFT and CWT analysis shows that chatter occurs at high frequency, beyond tooth passing frequency with high magnitudes. The feed rate was found to be the critical parameters which may drive the system into chaotic behavior. This was evidenced by exciting low and high frequencies well below and above the natural and harmonic frequencies. The frequency information contained in the wavelet transform describes the dynamic behavior of the system at the local time during the milling process. Thus, the CWT is recommended to use as a strategy not only for detecting the presence of chatter but also for an in-process control system.

Acknowledgments

This work was financially supported by the 'Center for Cyber-physical System Innovation' from The Featured Area Research Center Program within the framework of the Higher Education Sprout Project by the Ministry of Education (MOE) in Taiwan. Part of the funding also came from the Ministry of Science and Technology (MOST) in Taiwan under Grant no. MOST 107-2221-E-011-139-.

7. References

[1] Schmitz L, and Smith, S 2008 *Machining Dynamics: Frequency Response to Improved Productivity* (Springer Science & Business Media).
[2] Altintas Y 2012 *Manufacturing automation: Metal cutting mechanics, machine tool vibrations, and CNC design* (Cam-bridge university press).
[3] Tlusty, J. and Polacek, M 1963 *Proceedings of the International Research in Production Engineering Conference* (Pittsburgh, PA, ASME, New York) pp 454-465.
[4] Wei C, Liu, K and Huang H 2017 *Int J Auto Smart Tech* **6** (1) pp 25-36.
[5] Kundan K, Ramesh V and Kartik 2015 *Procedia Manuf* 1 pp 593-606.
[6] Dimla D and Lister P 2005 *Int J Mach Tool Manuf* **40** pp 739-768.
[7] Feng, J, Sun Z, Jiang Z, Yang L 2015 *Int J of Adv Manuf Tech* **82** pp 1909-1920.
[8] Emad Al-Regib and Jun Ni 2010 *J Dyna Syst, Measu Cont* **132** (3).
[9] Liu K, Tran Q, Qui Y and Chung C 2017 *Chatter Detection in Milling Process Based on Time-Frequency Analysis* (Proceedings of the ASME 2017 12th International Manufacturing Science and Engineering Conference, CA, US).
[10] Kunpeng Z, San Y and Soon G 2009 *Int J Mach Tool Manuf* **49** pp 537-553.
[11] Schmitz T, Medicus K and Dutterer B 2002 *Mach Sci Tech* **6** pp 207-225.
[12] Huang P, Li J, Sun J and Zhou J 2013 *Int J of Adv Manuf Tech* **64** pp 613-621.
[13] Berger S, Harlay J and Rokni M, Papadopoulos M 1988 *J Sound Vib* **213** (5) pp 813-827.
[14] Ma L, Melkote S and Castle J 2013 *J Manuf Sci Eng* **135**.
[15] Khraisheh K, Pezeshki C and Bayoumi E 1995 *J Sound Vib* **180** (1) pp 67-87.
[16] Michel Misiti, Yves Misiti, Georges Oppenheim and Jean-Michel Poggi 2002 *Wavelet Toolbox*.

IOP Conf. Series: Materials Science and Engineering **654** (2019) 012002 doi:10.1088/1757-899X/654/1/012002

Investigation of Aluminium and Copper Wound PMSM for Direct–drive Electric Vehicle Application

[1]Himavarsha Dhulipati, [1]Shruthi Mukundan, [2]Lucas Chauvin, [4]Christina Riczu [2]Afsaneh Edrisy, [3]Mark Kozdras, [4]Jennifer Bauman, [5]Saeid Habibi, [1]Jimi Tjong, and [1]Narayan C. Kar

[1]Department of Electrical and Computer Engineering, University of Windsor, Canada
[2]Department of Materials and Mechanical Engineering, University of Windsor,Canada
[3]CanmetMATERIALS, Natural Resources Canada
[4]Department of Electrical and Computer Engineering, McMaster University, Canada
[5]Department of Mechanical Engineering, McMaster University, Canada

dhulipah@uwindsor.ca

Abstract. This paper investigates aluminium and copper windings for a permanent magnet synchronous machine (PMSM) developed for direct–drive electric vehicle (EV) application. Previously, studies have been conducted on comparison of these windings in terms of thermal and electrical conductivity, cost and mass density. However, the impact of these windings on the machine's performance in terms of efficiency and torque has not been analysed. In this paper, for the same machine volume and geometry, a comparative analysis of PMSMs with copper and aluminium windings has been performed in terms of efficiency, torque, weight, operating speed range, ohmic losses and temperatures. Furthermore, as these machines are developed for direct–drive EV application, drive–cycle-based analysis was conducted for urban and highway cycles for a 2013 Ford Focus vehicle dynamics model. For these drive cycles, analysis in terms of torque speed characteristics and maximum energy density efficiency for both the machines has been performed.

1. Introduction

Conventionally, permanent magnet synchronous machines (PMSMs) are wound with copper conductors due to its high electrical conductivity. However, aluminium windings are being investigated as a potential candidate as they are cost and mass sensitive and offer higher thermal heat capacity when compared to copper conductors [1-4] which is well–suited for high torque density machine applications. In a typical electric vehicle (EV), power output from the motor is transmitted to the wheels through the gearbox, differential and drive shaft [5]. Such a transmission system is found to contribute to a power loss between 2% and 20% of the motor output power [6]. Thus, by removing the gear box, the motor to wheel efficiency can be improved resulting in a direct–drive configuration. Challenges in designing a direct–drive PMSM are to obtain high torque density along with reduced cost. Thus, for such an application where reduced cost and increased torque density are of prime importance, aluminium conductors appear to be a good candidate to explore to copper conductors. Previously, comparison of aluminium and copper conductors were performed based on mechanical characteristics for induction machines [1]. [2],[4] compares these windings for physical properties, cost, resistivity and mass. In [3], to overcome the disadvantage of low electrical conductivity in aluminium conductors, pre–compressed Litz wires are proposed. However, there is limited literature on the effect of copper and aluminium conductors on the machine's performance.

In this paper, two electromagnetic models of PMSM developed for direct-drive EV, one with copper conductors and another with aluminium conductors are compared in terms of torque–speed

Content from this work may be used under the terms of the Creative Commons Attribution 3.0 licence. Any further distribution of this work must maintain attribution to the author(s) and the title of the work, journal citation and DOI.
Published under licence by IOP Publishing Ltd

characteristics, efficiencies, ohmic losses and temperatures for different scenarios. Furthermore, the two machines with copper and aluminium windings have been analysed for urban and highway drive cycles to understand the feasibility of the machine for direct–drive application. Performance in terms of maximum energy density of these drive–cycles has been discussed.

2. Properties of Copper and Aluminium Windings

The physical properties of copper and aluminium windings used for this study are shown in Table 1 [7-8]. The basic data of the PMSM developed for the direct–drive EV application is shown in Table 2. For the same dimensions, conductor size, turns per slot and slot fill factor of 64%, two PMSMs, one with copper wires and another with aluminium wires were developed and analysed. Due to difference in resistivity of both the materials, the per–phase resistance of copper was found to be 2.1 Ω and for that of aluminium wound PMSM was 3.3 Ω. For the same geometry and material consumption, there was a 14.7% reduction in overall weight of the aluminium wound machine when compared to copper wound PMSM as shown in Table 3.

Table 1. Physical properties of aluminium and copper conductors at 20°C

Property	Copper	Aluminum
Resistivity [$\Omega mm^2/m$]	0.01	0.03
Mass Density [g/cm^3]	8.89	2.7
Thermal Conductivity [W/mK]	388	234
Tensile Strength [MPa]	248-276	62-96.5
Melting Point [°C]	1085	660

Table 2. Basic machine data

Outer Diameter [mm]	195
Active Length [mm]	75
Number of Slots/Poles	36/34
Base Speed [rpm]	405
Wire Diameter [mm]	1.02
Rated Torque [Nm]	70

Table 3. Parameters of Aluminium and Copper wound PMSM

Parameter	Copper	Aluminum
Slot Fill Factor [%]	64	64
Phase Resistance [Ω]	2.1	3.3
Active Weight [kg]	10.5	8.95
Winding Cost/kg [US$]	6.96	2.24

3. Comparison of copper and aluminium windings for a direct–drive PMSM

In this section, three scenarios are established for which the copper and aluminium wound PMSM machine's performance was analysed in terms of efficiency, torque, winding losses and temperature are analysed.

3.1 Case 1: Considering same output torque of 70 Nm in copper and aluminium wound PMSM

As previously discussed in section 2, the copper and aluminium wound PMSMs have the same geometry and conductor gauge. For a rated peak current of 9 A in the copper wound PMSM, the output torque at base speed of 405 rpm was found to be 70 Nm. To deliver the same torque for the aluminium wound PMSM, a rated peak current of 13 A was calculated for the aluminium wound PMSM based on the ratio of difference in resistivity of these two materials. These peak current values were used to analyse efficiency, torque and speed maps for the entire speed range which are illustrated in Figure 1.

Both the machines deliver a rated torque of 70 Nm until a speed of 405 rpm. However, to deliver the same torque in aluminium wound PMSM the rated current was increased by 1.4 times. Efficiency was calculated based on the mechanical output and losses. Losses in a PMSM are categorized as core, ohmic, mechanical and magnet loss. For direct–drive application the base speed is low when compared to high speed machine in traction application, therefore, speed related losses such core and mechanical loss are low, and the total machine loss is most dependant on by ohmic loss. Ohmic loss in any machine is dependent on square of current and resistance of the conductor used. In this study, the per phase ohmic losses are 156 W and 225 W in copper and aluminium wound PMSMs, respectively. Due to this difference in ohmic loss, the efficiency of aluminium wound PMSM was found to be 4.15% lower than the one in copper wound PMSM at rated speed and torque condition of 405 rpm and 70 Nm as shown in Figure 1(c).

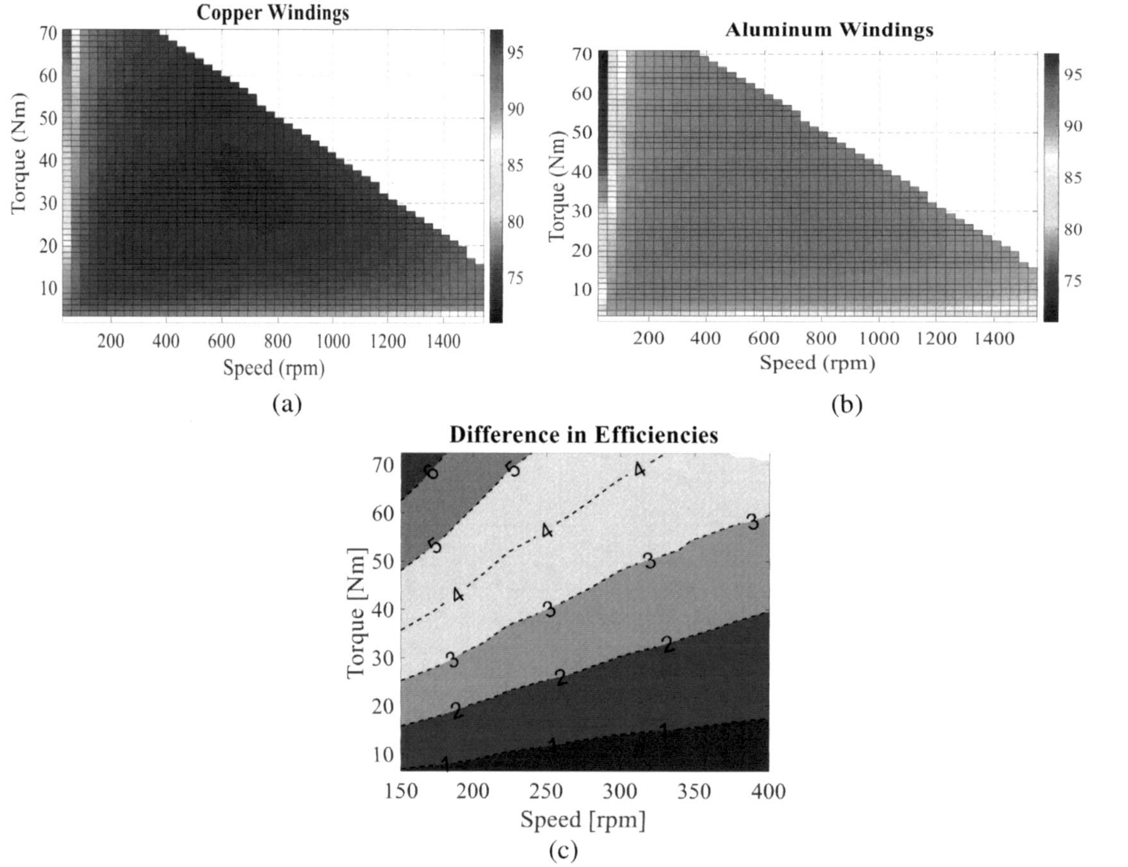

Figure 1. Efficiency maps of copper and aluminium wound PMSMs for entire torque–speed range. (a) Copper windings. (b) Aluminium windings. (c) Difference in Efficiencies until base speed of 405 rpm.

3.2. Case 2: Considering same ohmic loss of 156 W in copper and aluminium wound PMSM

In this scenario, for different per phase resistances of 2.1 Ω and 3.3 Ω in copper and aluminium wound PMSMs respectively, the current magnitudes were computed to obtain the same ohmic loss of 156 W per phase and output of 3 kW in both the machines. The resulting difference in output torque is shown in Figure 2(a). Aluminium wound PMSM delivered a torque of 58 Nm at base speed of 405 rpm whereas, copper wound PMSM delivered 70 Nm. This 18% reduction in torque production is due to lower conductivity of aluminium. Furthermore, torque–speed characteristics for both machines were analysed and are shown in Figure 2(b). Both the machines have maximum speed of 1,200 rpm. However, the base speed of aluminium wound PMSM increased by 20% when compared to copper wound PMSM, thereby offering wider constant torque region. This increase in base speed is attributed to reduced voltage and current in the aluminium wound PMSM.

3.3. Study of ohmic losses and efficiency as a function of temperature for copper and aluminium wound PMSMs

To deliver the same output torque of 70 Nm at a base speed of 405 rpm, copper and aluminium wound machines were analysed for ohmic losses and efficiencies at varying winding temperature. Resistance temperature coefficient of 0.00393 and 0.004 were considered for copper and aluminium wires, respectively [7]. Based on the temperature coefficients, the change in resistance from 20°C to 100°C was computed. Due to variation of resistances with temperature, corresponding magnitudes of currents were calculated to deliver 70 Nm torque for both the machines. The variation of ohmic loss and efficiency for aluminium and copper wound PMSMs as a function of temperature is shown in Figure 3. 3.8% and 2.65% reduction in efficiency were observed in aluminium and copper wound PMSMs. respectively. The larger reduction of efficiency in aluminium wound machine is attributed to reduced thermal conductivity of the material as seen in Table 2 when compared to copper windings.

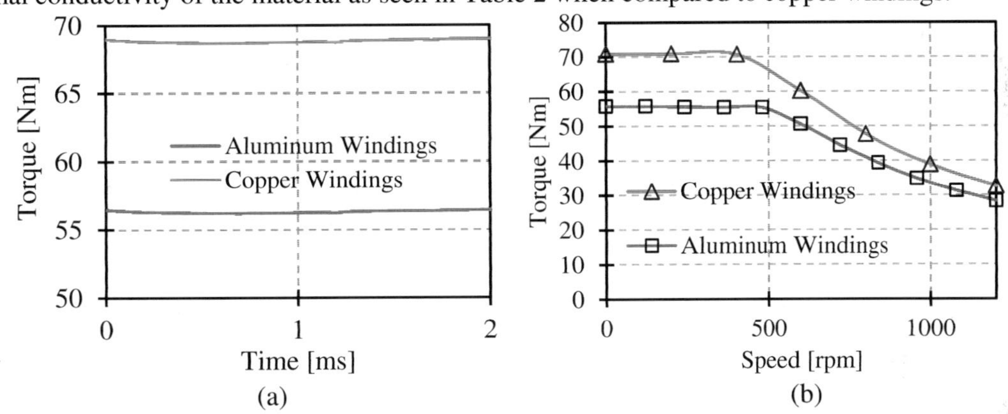

Figure 2. Comparison of torque profile for copper and aluminium wound PMSMs.(a) Torque comparison at rated speed of 405 rpm. (b) Torque–speed characteristics over entire operating range.

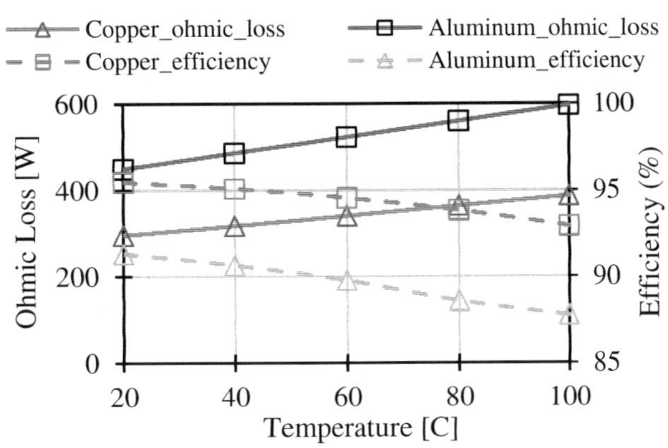

Figure 3. Ohmic losses and efficiency variation with temperature for copper and aluminium wound PMSMs.

IOP Conf. Series: Materials Science and Engineering **654** (2019) 012002 doi:10.1088/1757-899X/654/1/012002

4. Drive cycle–based performance analysis of copper and aluminium wound PMSMs

In this section, drive–cycle-based analysis is performed on the copper and aluminium wound PMSMs. The torque-speed characteristics of these PMSMs are analysed for case 1 scenario and efficiencies are analysed for case 2 scenario. This analysis simulates real–life driving conditions; specifically, the two most common drive-cycles which are UDDS (urban) and HWFET (highway). The characteristics used for analysis of these drive cycles are shown in Table 4. These drive cycles are variations of speed as a function of time which act as inputs to the vehicle dynamics model. Parameters of the vehicle dynamics model were selected for a 2018 Ford Focus vehicle [9]. Based on the speed variation for the aluminium wound PMSM, the torque demand is computed from the vehicle dynamics model using tractive forces acting on the vehicle. Three tractive forces were considered: aerodynamic drag, rolling resistance friction and force due to gravity. Since this is for a direct–drive EV, gear ratio was considered as unity, however a final drive ratio of 2.57 was considered for calculating the torque delivered by the motor [9].

For an urban drive cycle shown in Figure 4(a), the torque–speed points were computed using the vehicle dynamics model. These points were analysed for the torque–speed characteristics obtained from copper and aluminium wound PMSMs as shown in Figure 4(b). Even though the aluminium wound PMSM delivers a rated torque of 58 Nm, all of the torque speed points from the urban drive–cycle are satisfied by both the machines. Further, analysis of effect of drive–cycle points on efficiency maps were conducted. Figures 4(c) and (d) shows the urban drive cycle points on efficiency maps of copper and aluminium wound PMSMs respectively. A reduction of 1.17% efficiency was observed in aluminium wound PMSM when compared to copper wound machine at maximum energy density points of the urban drive–cycle.

Table. 4. Drive–cycle characteristics

	Urban	Highway
Average Speed (km/h)	34.1	77.7
Duration (sec)	1,874	765
Total Distance (km)	17.77	16.51

Similarly, for highway drive–cycle shown in Figure 5(a), the torque speed points were computed using vehicle dynamics model and were analysed for the torque –speed characteristics of both machines as shown in Figure 5(b). As seen from the figure, all the torque–speed points of highway drive–cycle are well within the characteristics of both the machines. Analysis of the highway drive cycle points on efficiencies of copper and aluminium wound machines are illustrated in Figures 5(c) and (d) respectively. A reduction of 1.7% efficiency was observed in aluminium wound PMSM when compared to copper wound machine at maximum energy density points for the highway drive–cycle.

5. Conclusion

Comparative analysis of copper and aluminium windings for direct–drive PMSM has been performed in this paper in terms of torque–speed characteristics, efficiencies, ohmic losses and temperatures. Analysis of both the machines for urban and highway drive–cycles was also conducted. Even though the efficiency of aluminium wound PMSM was approximately 1.7% less than copper wound PMSM for the two drive– cycles at maximum energy density points, it satisfies all the speed and torque load points offering 14.7% less motor weight and 67% less winding cost per kg when compared to copper wound PMSM. This difference in efficiency between both the machines is significant as they were designed for low–speed application where the predominant losses are from the windings. However, there is a scope of implementing these windings for high–speed application, where winding losses are low and have less impact on efficiency.

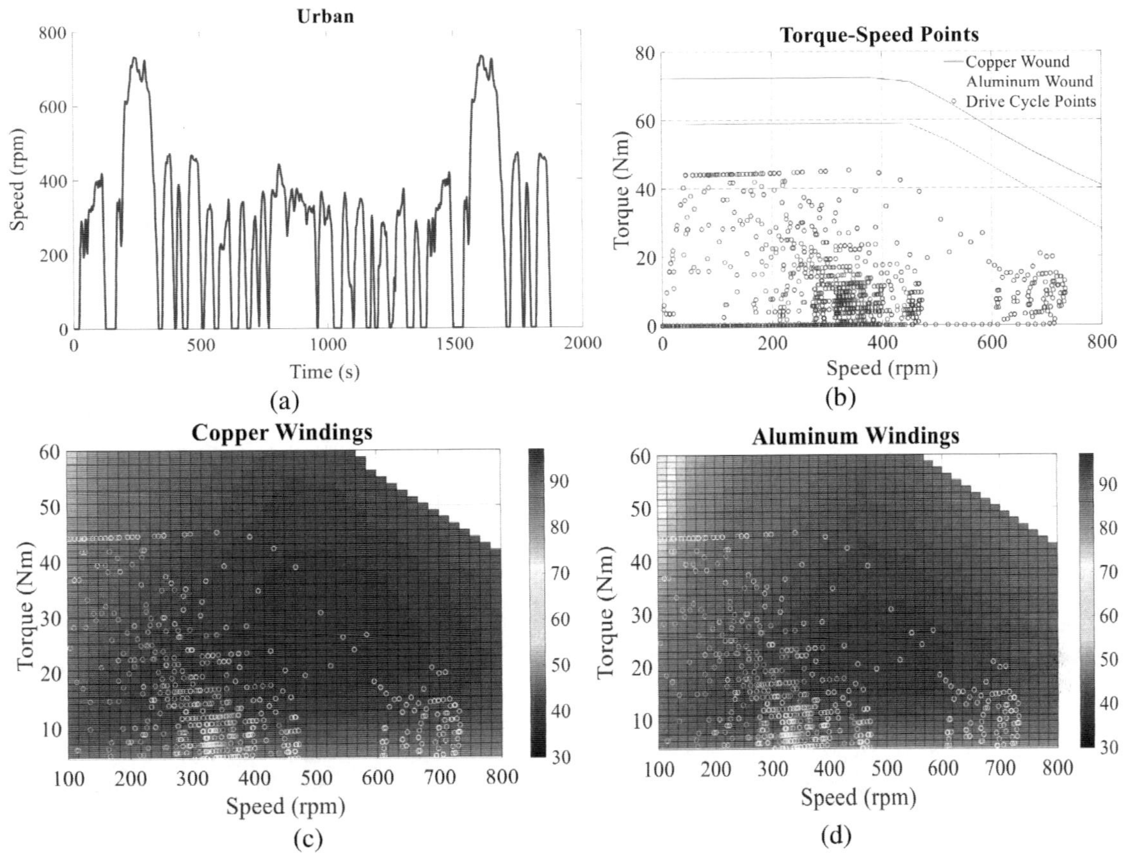

Figure 4. Urban drive–cycle performance characteristics. (a) Urban drive cycle. (b) Torque–speed load points. (c) Torque speed points of urban drive–cycle on efficiency map of copper wound PMSM.(d) Torque speed points of urban drive–cycle on efficiency map of aluminium wound PMSM.

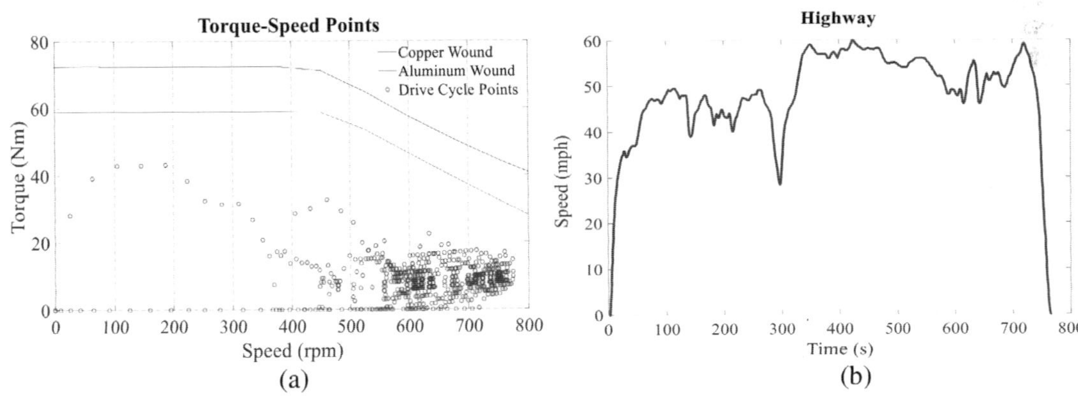

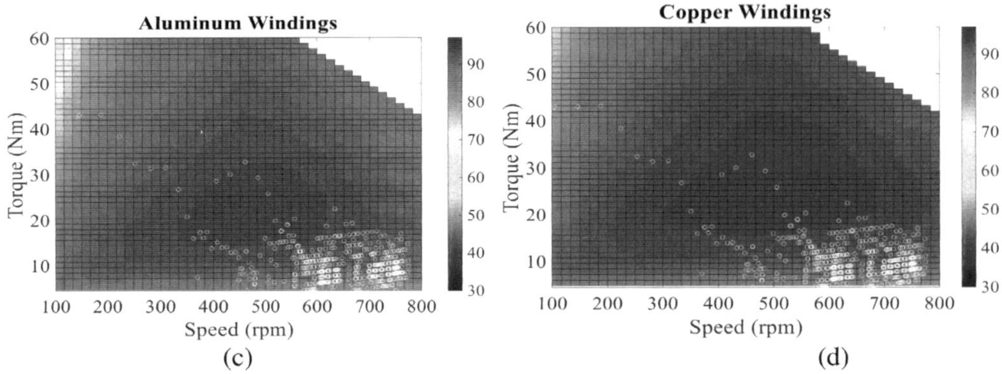

Figure 5. Highway drive cycle performance characteristics. (a) Highway drive–cycle. (b) Torque–speed load points. (c) Torque speed points of highway drive–cycle on efficiency map of copper wound PMSM.(d) Torque speed points of highway drive–cycle on efficiency map of aluminium wound PMSM.

6. References

[1]M. Iorgulescu, "Study of single–phase induction motor with aluminum versus copper stator winding," in proc. of *International Conference on Applied and Theoretical Electricity*, Craiova, pp. 1–5, 2016.

[2]C. R. Sullivan, "Aluminum windings and other strategies for high–frequency magnetics design in an era of high copper and energy costs," *IEEE Transactions on Power Electronics*, Vol.23, pp. 2044–51, 2008.

[3]J. D. Widmer, R. Martin and B. C. Mecrow, "Pre–compressed and stranded aluminum motor windings for traction motors," *IEEE Transactions on Industry Applications, Vol 52, No.3, May 2016.*

[4]R. Wrobel, D. Salt, N. Simpson and P. H. Mellor, "Comparative study of copper and aluminum conductors - Future cost-effective PM machines," in proc. of *IET International Conference on Power Electronics, Machines and Drives*, Manchester, 2014, pp. 1-6.

[5]K. L. V. Iyer, S. Mukundan, H. Dhulipati, K. Mukherjee, B. Minaker, and N. C. Kar, "Design considerations for permanent magnet machine drives for direct–drive electric vehicles." in proc. of *Electric Machines Drives Conference*, pp. 1170–1176, 2015.

[6]BorgWarner Drivetrain Systems, 2011, Available: http://www.evwest.com/support/borgwarneredrive.pdf

[7]Superior Essex Data Sheet, Available: https://www.superioressex.com/uploadedFiles/Magnet_Wire_and_Distribution/North_America/Magnet_Wire_–_Winding_Wire/EngData_book_linked.pdf

[8]ASM Handbook, Available: http://sme.vimaru.edu.vn/sites/sme.vimaru.edu.vn/files/volume_2 properties_and_selection_nonf.pdf

[9]2018 Ford Focus Electric Specifications, [online], Available: https://www.ford.ca/cars/focus/models/focus–electric

IOP Conf. Series: Materials Science and Engineering **654** (2019) 012003 doi:10.1088/1757-899X/654/1/012003

Health Condition Evaluation of Servo Turret Based on Time Sequence Signal Feature Map

W Z Chen[1,2], F Chen[1,2], T Ma[1,2] and B B Xu[1,2]

[1] Key Laboratory of CNC Equipment Reliability, Ministry of Education, P.R.China
[2] School of Mechanical and Aerospace Engineering, Jilin University, Changchun, P.R.China

Corresponding author: chenfeicn@jlu.edu.cn

Abstract. The health condition of servo turret depends machining quality of CNC lathe. Performance tests of servo turret are taken to evaluate the health condition and overall status during its service life. This paper selects the core signals during one tool change-cutting cycle from multichannel physical signals and describes the performance of tool change-cutting cycle by defining a series of feature matrixes. In order to clearly and visually express the health condition of different working stages and tool position, a multidimensional features map is designed according to the combination of different time sequence feature matrixes. The above method is applied on a servo turret test platform which is equipped to acquire abundant signals. The result shows that the feature map clearly defines the health condition with one easily interpreted picture instead of unarranged signals. The proposed method provides a new idea of feature selection and visual expression, which can be an instructive tool for health evaluation and fault isolation.

1. Introduction

As core components in CNC lathe, the performance of servo turret determines the machining accuracy of the workpiece. With the development of manufacturing technology, higher performance and reliability of servo turret are required. Regular health assessment of the servo turret is meaningful for engineers to assess the performance and arrange maintenance policy quickly. This, in turn, decreases the production downtime while minimizing the impact of failures due to malfunctioning [1-2].

Abundant information appears during the working period of servo turret, including tool position, load, vibration, current, precision, deformation. There is an ever-increasing interest in multi-disciplinary research on multi-sensor data fusion technology, driven by its versatility and diverse areas of application. Therefore, it is a real need for an analytical review of recent developments in the data fusion domain [3]. Ming Dong and David proposed a multi-sensor integration platform based on the semi-hidden Markova model, which realized multi-information fusion by adjusting the importance assigned to each sensor through discriminant function analysis [4]. Using Dempster-Shafer evidence theory, Otman Basir established a fusion model including engine vibration, temperature, sound, pressure, and other sensor signals, and effectively diagnosed engine faults [5]. H.O.A. Ahmed et al. proposed a method based on sparse feature coding to monitor the health status of the bearing, equipping the system with the ability to process data and feature during collecting data [6].

The health assessment system of servo turret has not been fully automated, and the applying of visualization result to engineers can help them make decisions more efficiently and accurately. In Lee J's review of visual assessment methods for PHM, degradation charts, performance radar charts, failure charts, and risk radar charts are described [7]. However, the visual information interaction for

Content from this work may be used under the terms of the Creative Commons Attribution 3.0 licence. Any further distribution of this work must maintain attribution to the author(s) and the title of the work, journal citation and DOI.
Published under licence by IOP Publishing Ltd

different fault modes, positions, and fault effects is still one of the issues in the field of fault diagnosis of complex electromechanical products. When the information is transmitted to the receiver in a pattern with visual laws, the human brain can more efficiently understand [8] and make further decisions. Yujie Xu et al. obtained the transfer rule of characteristic caused by performance recession through the engine air path map. Based on the analysis of the structure of the characteristic space in different environments and working conditions, fault type, and position of the engine can be identified quickly [9]. Therefore, it is of great practical value to establish the correlation of feature of servo turret in different working stages, and design a visualized health analysis tool for the health evaluation.

In order to analyze, extract and visualize the expression of features in different stages of servo turret, this paper classifies signals into different stages after which a signal state map that could quickly identify unnatural links and abnormal positions of the servo turret was designed. This map provides a new idea for the abnormal isolation and fault diagnosis of the work cycle. On the other hand, the proposed method optimizes the relationship of multi-dimensional signals of mechanical equipment, facilitates the management of data, and brings a new idea for mining the information of mechanical equipment.

2. Signals in the different working stage

Servo turret is responsible for replacing cutters so that the CNC lathe can perform different cutting types. During a work cycle, there is a conversion between the electric energy and the tools' kinetic energy, resulting in vibration and current signal, as shown in Figure 1. During the tool change process, the vibration and current signals are used to characterize the performance of the servo turret. The repeated positioning accuracy after the tool change is completed to characterize the precision performance of the servo turret, and the vibration and deformation generated by cutting force under the process of cutting can fully reflect the affordable loading ability of the servo turret. The increasing of motor current and vibration signal of the A period in Figure 1 reflects the abnormal condition of the motor and the transmission system of servo turret. The main reason for the increasing of locking vibration and the poor repeatability of the B period in Figure 1 is a sign that the meshing-toothed discs are degrading. Large deformation and severe vibration in the C period indicates that the rigidity and dynamic performance of the servo turret are defective. Multi-sensor information fusion technology based on a variety of sensing signals that can accurately and completely reflect the working state of the CNC lathe and has a positive impact on the aspects of subsequent maintenance and fault diagnosis.

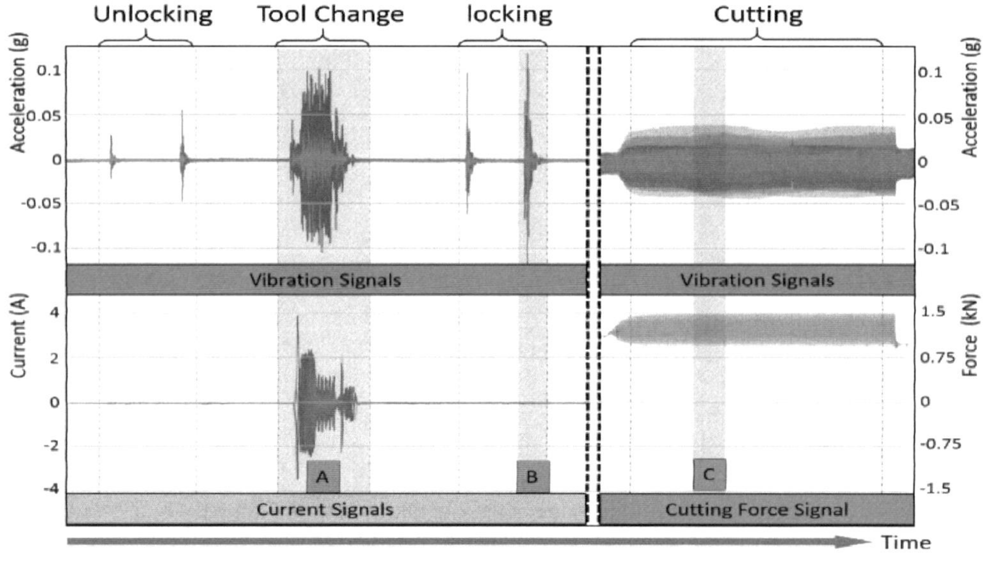

Figure 1. Signals during tool change and cutting cycle.

3. Feature defining of different working stages

3.1. Feature selection at the transposition stage

The process of transposition stage can be evaluated by the effective value of current I, starting vibration V_a and tool change vibration V_b. Let the performance index characteristic matrix from x position to y position be $T_{(x,y)}$, the feature quantity matrix in the tool transposition process can be expressed as:

$$T_{(x,y)} = [I\ V_a\ V_b] \tag{1}$$

Suppose that the servo turret has N positions, the tool transposition state matrix HI_1 switched between different tools position can be described as:

$$HI_1 = \begin{bmatrix} T_{(1,2)} & \cdots & T_{(1,N)} \\ \vdots & \ddots & \vdots \\ T_{(N,1)} & \cdots & T_{(N,N-1)} \end{bmatrix} \tag{2}$$

3.2. Feature selection at locking stage

After the tool rotates to the target position, the double-toothed fluted disc mesh with the movable fluted disc connected to the cutter head and the located fluted disc fixed to the servo turret enclosure, which further ensures the location accuracy and strengthens the rigidity of servo turret. The locking vibration of the fluted disc is related to the locking force generated by hydraulic pressure and the contact quality among fluted discs. Since the oil pressure value that controls the locking force is adjustable, this paper selects the RMS of locking vibration $V_{c(x,y)}$ under the recommended oil pressure for evaluation:

$$HI_2 = \begin{bmatrix} V_{(1,2)} & \cdots & V_{(1,N)} \\ \vdots & \ddots & \vdots \\ V_{(N,1)} & \cdots & V_{(N,N-1)} \end{bmatrix} \tag{3}$$

Repeated positioning accuracy is used as the accuracy evaluation index for repositioning the machining point after the tool change, which directly affects the machining error. Therefore, the performance index needs to be tested regularly and defined as HI_3:

$$HI_3 = [P_1, P_2, \cdots, P_N]' \tag{4}$$

3.3. Feature selection at cutting stage

The rigidity of the servo turret system directly affects the cutting vibration and then affects the roughness and precision of the workpiece. The failure of the servo turret system often causes abnormal vibration. Therefore the cutting vibration under specific loading force is used to evaluate the performance of the servo turret during cutting indirectly. In order to simplify the test, the cutting vibration under a typical working condition can be monitored according to the cutting load spectrum. According to the vibration test standard, vibration is evaluated using the root mean square value.

$$x_{rms} = \sqrt{\frac{1}{N}\sum_{i=1}^{N} x_i^2}\ , i = 1,2,\cdots,N \tag{5}$$

According to the relationship between the loading force and the deformation of the cutting tool, the least square fitting is performed to solve the rigidity k value of the servo turret.
Hence, the characteristic matrix for the cutting vibration of different tool positions can be expressed as:

$$HI_4 = [x_1\ k_1, x_2\ k_2, \cdots x_N\ k_N]' \tag{6}$$

4. Health feature map

For those feature index that already has the industry standard, the degree of the deviation from the factory standard can be directly used for condition monitoring, such as repeated positioning accuracy, cutting vibration and servo turret rigidity. Others need to be artificially defined, combining with the health status data. In order to comprehensively consider features in different working stages of the servo turret and locate the weak link of servo turret, a visual map is established in this part.

As is shown in Figure 2, signals in one tool change cycle of servo turret are selected first while some coral features are extracted. In order to explore the condition of different position, features are classified into start position matrix, arrived position matrix, and unrelated position features. These features are arranged by time sequence and are designing several noteworthy dimension of equipment. Health condition map is designed to vividly exam and evaluate the health condition of servo turret. When an abnormal situation happens, related operation data can be used to assist health evaluation and abnormal diagnosis.

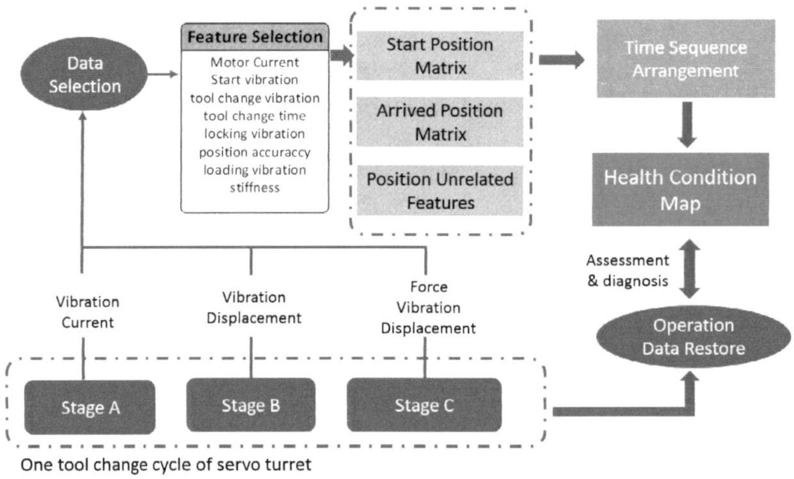

Figure 2. The framework of the proposed method.

The feature maps established in this paper include tool position, working stage, features, and health level. The dimensions of the working stage and features are based on the characteristic signals generated by different stages and different parts of the full-cycle servo turret. The health level is scored according to the degree of deviation between the characteristic value and the set value and is converted into RGB values according to the score value, which has three levels of excellent (green), good (orange) and poor (red). The feature map has two modes: relative reference and absolute reference. In the relative reference, the map health level describes the deviation of each feature quantity from the initial value of servo turret itself. The value of the feature value directly describes the absolute reference without comparison.

The health matrix is arranged in the way of working stage of servo turret, and is graphically converted into a full-cycle performance mapping framework as shown in Figure 3:

$$CHI = \begin{bmatrix} T_{(1,2)} \cdots T_{(1,N)} \\ \vdots \quad \ddots \quad \vdots \\ T_{(N,1)} \cdots T_{(N,N-1)} \end{bmatrix} \begin{bmatrix} V_{(1,2)} \cdots V_{(1,N)} \\ \vdots \quad \ddots \quad \vdots \\ V_{(N,1)} \cdots V_{(N,N-1)} \end{bmatrix} \begin{bmatrix} P_1 \\ \vdots \\ P_N \end{bmatrix} \begin{bmatrix} x_1 \ k_1 \\ \vdots \ \ \vdots \\ x_N \ k_N \end{bmatrix} \tag{7}$$

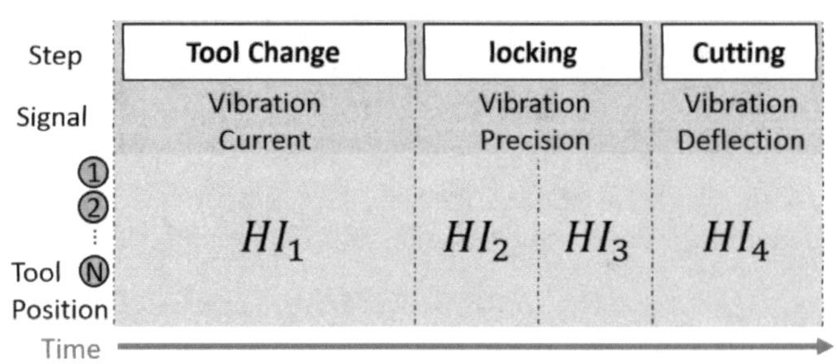

Figure 3. Designed feature map.

IOP Conf. Series: Materials Science and Engineering **654** (2019) 012003 doi:10.1088/1757-899X/654/1/012003

5. Experiment and application

In order to verify the proposed method, an experiment in servo tool test system is performed, and health performance is evaluated by using the feature map. Servo turret controlled by PLC can change tool automatically while the cutting force is loaded by electro-hydraulic servo loading device. In the test system, vibration sensors are arranged on the shell, current sensors are connected to the three-phase servo motor, the force sensor is installed on the end of the force loading mechanism, and laser displacement sensor is used to detect the repetitive positioning and deformation after loading. All detected signals are captured by a data acquisition device which provides a basis for health assessment.

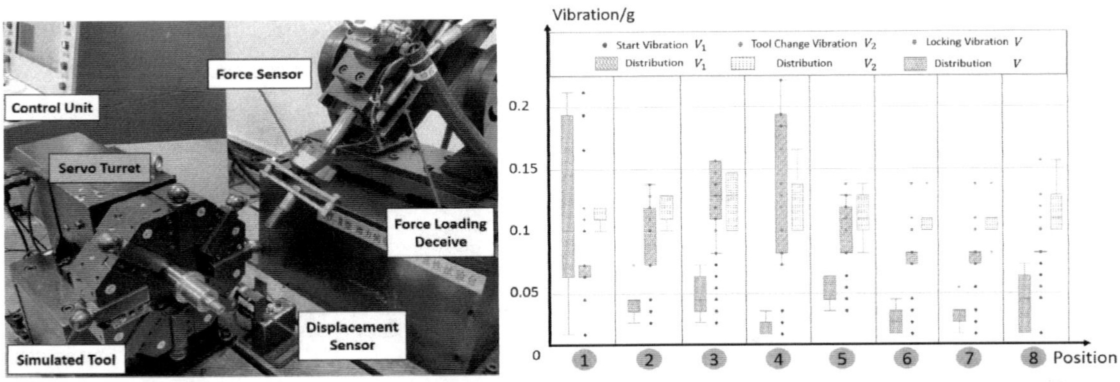

Figure 4. Test platform of servo turret and vibration data of different stages.

The average value of RMS of current turning to the same target position varies from 1.39A to 1.42A with maximum variance for the difference between different interval transpositions is also only 0.06. Therefore, it can be considered that the current is an inherent characteristic of the servo motor, which is independent of tool position. The relationship between starting vibration, tool change vibration, and locking vibration of different tool positions are shown in Figure 4. It can be found that the difference in starting vibration of different tool positions is noticeable, and the maximum value is six times the minimum value. The average value of the tool change vibration is similar with small variance. So the relationship between the tool change vibration and the tool position can be ignored. The locking vibrations under different target positions are different, with the maximum value is 2.5 times as the minimum value.

The test results of the repeated positioning accuracy P_n and the loading vibration x_n of different tool positions are shown in Table 1. The repeating positioning accuracy is the average value of the test results with each position rotating forward and backward 3 times. The loading vibration is measured under a typical load spectrum condition, where the cutting force is 300N and the frequency is 28Hz.

Table 1. Features of performance

NO.	1	2	3	4	5	6	7	8
P_n	2.3	5.9	3.5	2.5	3.4	4.5	1.1	4.0
x_n	0.0032	0.0018	0.0029	0.0014	0.0015	0.0028	0.0028	0.0025
k_n	15.6	13.3	14.5	14.4	13.6	14.1	17.5	15.6

The deviation area of the alarm can be defined according to actual demand and severity. The default setting is 50% off average data for exceptions. For different stages, the threshold of excellent, functional, and bad can be customized or be in a ratio of 3:4:3. Combined with the test data and analysis results, the main characteristic indicators, including the starting current, starting vibration, processing vibration, locking vibration, repeat positioning accuracy, loading vibration, and stiffness are adjusted. Among the indicators, current characteristics, turning vibration, repeated positioning accuracy and loading vibration are independent of the tool position, starting vibration can identify the initial tool position and locking vibration identifies the target tool position. Then the CHI′ matrix can be constructed as follows:

$$\mathrm{CHI}' = [C_{b1}\ V_{a(x,y)}\ V_b\ V_{c(x,y)}\ P_n\ x_n\ k_n] \quad (8)$$

The health matrix is established in combination with the test data of the above full cycle and converted into a feature map by the relative reference, as shown in Figure 5. It can be seen from the figure that the first and position 2 have a large starting vibration, but the vibration is small when the tool is locked. The locking vibration of the third and fourth position is severe and should be concerned. The repeat positioning accuracy of the second and sixth position is the worst, and the first and the third tool positions vibrate significantly in the actual cutting. The seventh and eighth position has the worst stiffness.

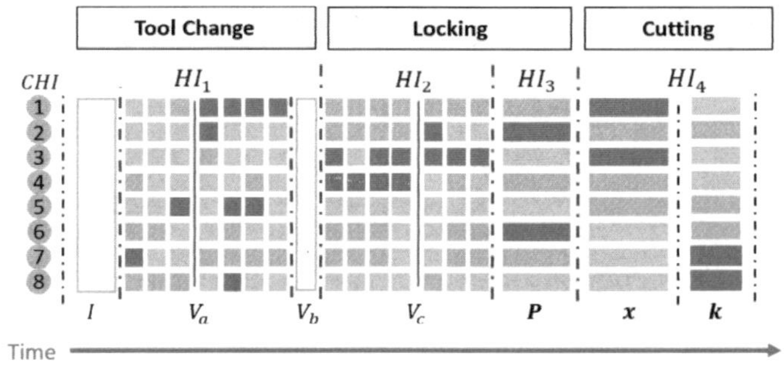

Figure 5. Health condition evaluation result of tested servo turret.

6. Conclusion

Feature extraction is performed in this paper to quantify describe sensitive information of servo turret. A four-dimensional feature map is designed so that the health condition of servo turret can be visually and fully explained. Since experts focus on different aspects like reliability, overall performance, precision, and fault diagnosis, the proposed feature map is unable to express every useful information for users fully. However, the method classified the typical signals and its features and further rearrange them into a physical and time-related map, which provides a useful guide for health evaluation and even maintenance policy. The method describes in this paper has application wildly in the intelligent health evaluation and condition motoring of complicated mechanical equipment.

Acknowledgments

This paper is supported by the National Science and Technology Major Project (2018ZX04006039) and Program for JLU Science and Technology Innovative Research Team (2017TD23).

7. References

[1]Alaswad S Z and Xiang Y S, 2017 A review on condition-based maintenance optimization models for stochastically deteriorating system, *Reliability Engineering and System Safety* **157** 54–63.

[2]Yin S, Ding S X, Xie X C and Luo H, 2014 A review on basic data-driven approaches for industrial process monitoring *IEEE TRANSACTIONS Transactions on industrial electronic* **61** 6418-28.

[3]Khaleghi B, Khamis A, Karray F O and Razavi S N 2013 Multi-sensor data fusion: A review of the state-of-the-art *Information Fusion* **14** 28–44.

[4]Dong M and He D 2007 Hidden semi-Markov model-based methodology for multi-sensor equipment health diagnosis and prognosis *European Journal of Operational Research* **178** 858-78.

[5]Basir O and Yuan X H 2007 Engine fault diagnosis based on multi-sensor information fusion using Dempster-Shafer evidence theory *Information Fusion* **8** 379-86.

[6]Ahmed H O A, Wong M L D and Nandi A K 2018 Intelligent condition monitoring method for bearing faults from highly compressed measurements using sparse over-complete features *Mechanical Systems and Signal Processing* **99** 459-77.

[7]Lee J, Wu F J, Zhao W Y, Ghaffari M, Liao L X and Siegel D 2014 Prognostics and health management design for rotary machinery systems—Reviews, methodology and applications *Mechanical Systems & Signal Processing* **42** 314-34, .

[8]Phil Simom 2014 *The Visual Organization: Data Visualization, Big Data, and the Quest for Better Decisions* (New Jersey: John Wiley & Sons Ltd)

[9]Hou X, Li W W and Zheng N 2018 Aero-Engine Gas-path Fault Diagnosis Based on Spatial Structural Characteristics of QAR Data 2018 *Annual Reliability and Maintainability Symposium (RAMS)* Reno **2018** 1-7.

IOP Conf. Series: Materials Science and Engineering **654** (2019) 012004 doi:10.1088/1757-899X/654/1/012004

Performance Analysis of Non–Oriented Electrical Steel with Optimum Texture for High–Speed Traction Motors

Shruthi Mukundan[1*], Mehdi Mehdi[2,3], Himavarsha Dhulipati[1], Lucas Chauvin[2], Afsaneh Edrisy[2], Youliang He[3], Jimi Tjong[1] and Narayan C. Kar[1]

[1]Department of Electrical and Computer Engineering, University of Windsor, Ontario, Canada
[2]Department of Mechanical, Automotive and Materials Engineering, University of Windsor, Ontario, Canada
[3]CanmetMATERIALS, Natural Resources Canada, Hamilton, Ontario, Canada

mukunda@uwindsor.ca

Abstract. The magnetic properties of non–oriented electrical steel (NOES) vary significantly with respect to the microstructure and crystallographic texture of the final steel sheets, which, in turn, are highly dependent upon the thermomechanical processing parameters used during hot rolling, cold rolling and annealing. This paper performs an exploratory performance analysis of NOES for use in high–speed traction motors, emphasizing the importance of obtaining an appropriate crystallographic texture to achieve the desired magnetic properties by controlling the annealing temperature and holding time. A 3.2% Si NOES annealed at 860ºC for 24 hours after hot rolling can result in reduced core losses after cold rolling and final annealing. This material was chosen for the performance analysis of a laboratory scale high–speed, high–power traction motor (45 kW, 10,000 rpm) using finite element analysis (FEA). The motor using the NOES is compared to that using a commercially available grain–oriented electrical steel (GOES) to highlight the feasibility and advantages of NOES for high–speed motor applications. In addition, the motor performance using the NOES is compared to that using commercial NOES to understand the scope of improvement obtained by optimizing the microstructure and texture of the steel sheet.

1. Introduction

Soft magnetic core (electrical steel lamination) used in electric motors requires optimum magnetic properties, e.g. high permeability and low core loss, to render high performance of the motor. Non–oriented electrical steel (NOES) is widely used in electric machines, and the thermomechanical processing steps involved in steel sheet production significantly affect the microstructure/texture and the magnetic properties [1–3]. This paper performs an exploratory performance analysis of a laboratory produced NOES with optimum microstructure and texture [4] for application in high–speed traction motors. The electrical steel sheets were manufactured using conventional rolling and annealing processes but analysed at different annealing temperatures and holding times to understand the texture variation, and to obtain optimum magnetic properties. Based on the analysis, a 3.2% Si NOES after final annealing at 850ºC for 60 minutes was chosen for motor performance evaluation. An electromagnetic model of a 45 kW, 10,000 rpm laboratory scale high–speed, high–power motor was developed and used for analysis. First, the NOES is compared with a commercially available grain–oriented electrical steel (GOES) to understand its suitability and advantages over GOES for motor

Content from this work may be used under the terms of the Creative Commons Attribution 3.0 licence. Any further distribution of this work must maintain attribution to the author(s) and the title of the work, journal citation and DOI.
Published under licence by IOP Publishing Ltd

applications. Motor performance characteristics in terms of rated torque production, maximum efficiency, rated core losses and operating speed range were obtained using finite element analysis (FEA). Furthermore, to understand the scope of improvement obtained from the optimized microstructure and texture, the motor performance using the NOES was compared with that using a commercially available NOES.

2. Non–Oriented Electrical Steel Processing and Analysis

Conventional manufacturing processes, i.e. hot rolling, annealing, cold rolling, and final annealing, were applied to the 3.2% Si NOES [4]. The steel was melted and cast into ingots, which were then reheated to 1050°C and hot rolled to a thickness of 20 mm in a reversing rolling mill in six passes. Further, to reduce the thickness to ~3.3 mm, a second hot rolling was applied under the same conditions. In order to remove surface oxides, the steel plates were then pickled in a hydrochloric acid solution. The hot–rolled plates were subsequently annealed at 860°C for 24 hours in argon–protected atmosphere and furnace cooled. The annealed plates were cold rolled to a final thickness of 0.35 mm and subsequently annealed again to obtain the final steel sheets. Final annealing was performed at various temperatures for different times to control the grain size and texture of the samples.

The textures of the steel were characterized using electron backscatter diffraction (EBSD) techniques in a field emission gun scanning electron microscope equipped with an EDAX Orientation Imaging Microscope (OIM) system. Using the grain orientation data, the grain sizes of the sample were calculated with the OIM software. The cross–section EBSD inverse pole figure (IPF) map of the final steel samples at different annealing temperatures and holding times and their corresponding textures are illustrated in Figure 1. AC and DC magnetization tests were performed on the final annealed steel strips using Epstein frame method to obtain the magnetic properties. It was found that the texture of the 3.2% Si NOES was optimized, i.e. a strong magnetically favourable θ–fibre (<001>//ND) was obtained, through a combination of a final annealing temperature of 750°C and a holding time of 60 minutes. This is because, annealing temperatures higher than 750°C resulted in preferred nucleation at shear bands, micro bands and fragmented grains which in turn increases the magnetically favourable θ–fibre and suppresses the magnetically detrimental γ–fibre. The steel strips after annealing under these conditions showed reduced DC core loss, the lowest AC core loss, and the maximum relative permeability. Thus, in this paper, the 3.2% Si NOES sample annealed at 850°C for 60 minutes is chosen for analysis.

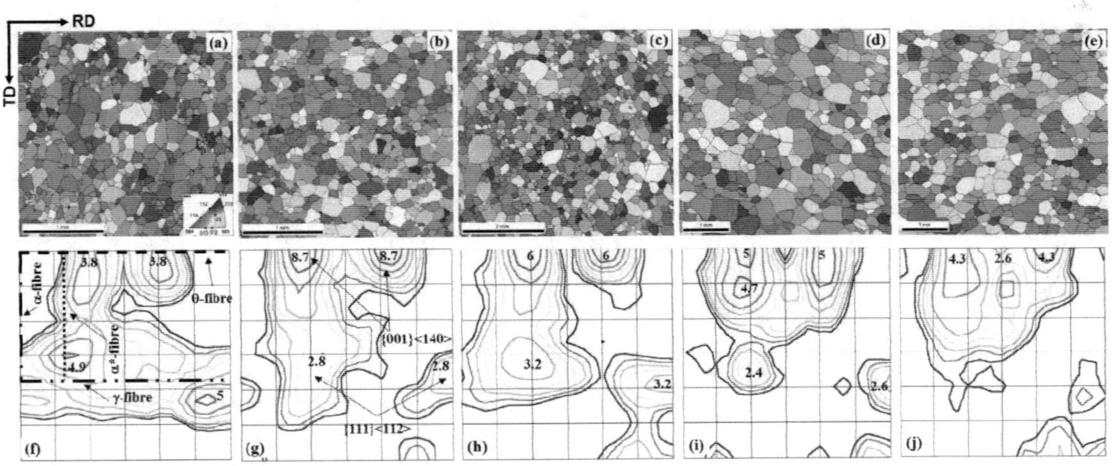

Figure 1. EBSD IPF maps showing the microstructure and texture of the final annealed steel sheet after final annealing at different temperatures and holding times: (a) 750°C for 180 minutes, (b) 850°C for 60 minutes, (c) 950°C for 30 minutes, (d) 1050°C for 10 minutes, (e) 1150°C for 2 minutes, (f) – (j) the corresponding textures [4].

IOP Conf. Series: Materials Science and Engineering **654** (2019) 012004 doi:10.1088/1757-899X/654/1/012004

3. Structural and Target Performance Parameters of the High–speed Traction Motor

In order to emphasize the inherent advantages provided by the NOES material, a high–speed traction motor was chosen for analysis. A permanent magnet synchronous machine (PMSM) with the structural and target performance parameters defined in Table 1 was modelled using finite element method incorporating with different steel core materials for analysis. The electromagnetic model of the motor is shown in Figure 2. The net copper area, permanent magnet volume and the machine dimensions are kept constant for analysis. A rated current of 78 A rms/phase was applied to the machine and the performance characteristics thus obtained were analyzed in terms of rated torque production, maximum efficiency, rated core losses and operating speed range.

Table 1. Structural and target performance parameters of the high–speed traction motor

Parameter	Value	Parameter	Value
Continuous Power	22 kW	Stator Outer Diameter	195 mm
Continuous Torque	70 Nm	Stack Length	75 mm
Peak Power	45 kW	Shaft Diameter	100 mm
Peak Torque	150 Nm	Air gap Length	0.5 mm
Rated Speed	3,000 rpm	Magnet Material	NdFeB 35
Target Efficiency	> 95%	Wire Diameter	8.5 AWG
Rated Current	78 A	Total Magnet Weight	0.53 kg
DC Bus Voltage	400 V	Total Steel Weight	9.08 kg

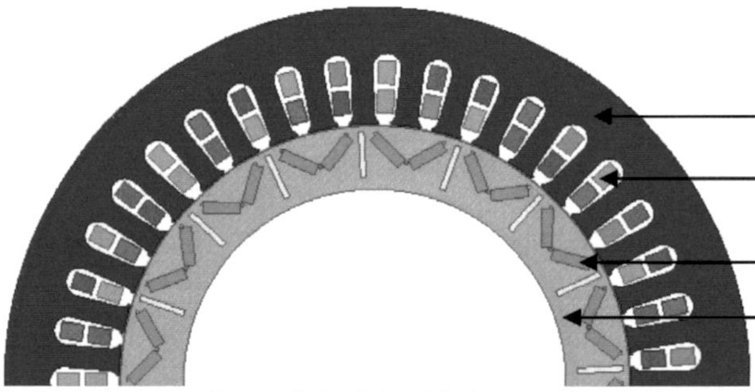

Figure 2. Developed electromagnetic model of the high–speed traction motor used for analysis.

4. Comparative Performance Analysis of High–Speed Motor: The Studied NOES vs. GOES

The magnetic properties of the studied NOES are compared to that of a commercially available GOES as shown in Figure 3. Both the electrical steels under consideration have 3.2% of silicon and the thickness of each lamination sheet is approximately 0.35 mm. It can be observed from Figure 3(a), that for flux densities below 1.25 T, the core losses of the studied NOES are comparable with that of the commercial GOES along the rolling direction (RD), and for higher flux densities, the core losses are slightly higher for the NOES. Further, the variation of core losses with frequency under a fixed magnetic flux density of 1 T is illustrated in Figure 3(b), and it can be seen that the losses for both the electrical steels are very similar.

For the above–mentioned electrical steel properties, the electromagnetic model of the motor indicated in Figure 2 is analyzed at a rated current of 78 A rms/phase and a rated frequency of 400 Hz. It can be observed from Figure 4(a), that both the machines have the same torque–speed and power–speed characteristics since the NOES and GOES materials have very similar magnetization curves. However, the maximum torque obtained was 20.6 Nm lesser than the target due to limited flux density capabilities of the steel materials. Hence, for both NOES and GOES, the magnetization properties have to be improved to obtain the rated torque performance, or increasing the rated current of the motor would result in the required torque production. However, in such case, the motor copper losses will increase significantly, reducing the rated motor efficiency. Thus, further studies are required to analyze the most suitable solution for both NOES and GOES materials.

Furthermore, it can be seen from Figures 4(b) and (c), that the motor with the studied NOES results in 48.5% lower rated core losses than that with commercial GOES material which results in a 0.45% increase in the rated motor efficiency. This can be attributed to the fact that the NOES sample under

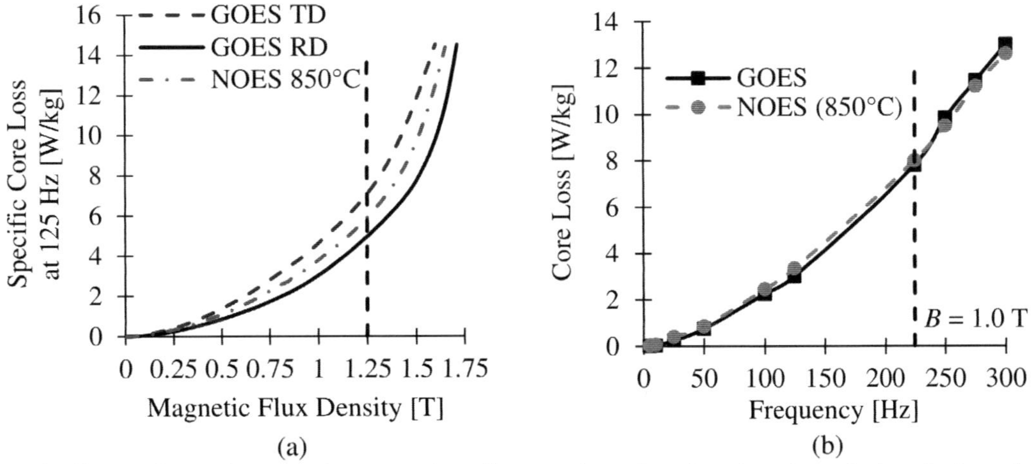

Figure 3. Comparison of magnetic properties of non–oriented and grain oriented electrical steels with 3.2% Si. (a) Core losses at different magnetic flux densities (with a fixed frequency of 125 Hz). (b) Core losses at various frequencies (with a fixed magnetic flux density of 1 T) [4].

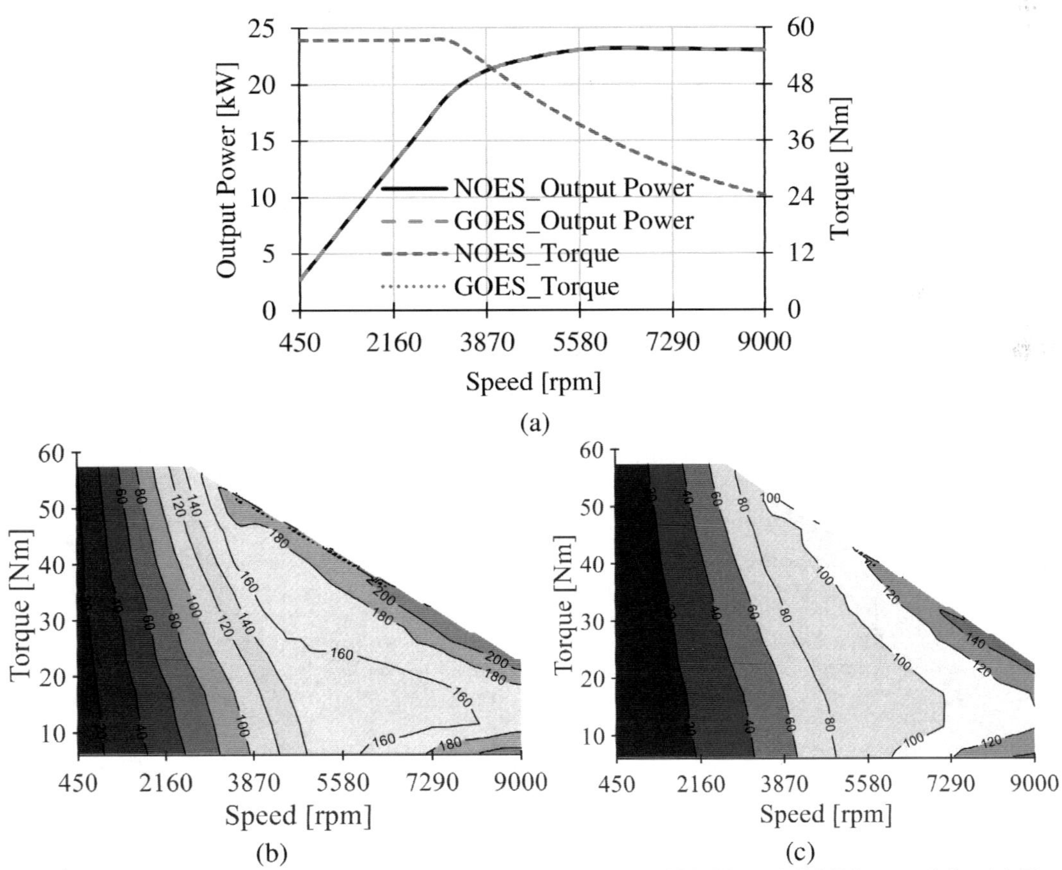

Figure 4. High–speed motor performance characteristics with NOES and GOES materials. (a) Torque and power speed characteristics. (b) Core loss map of motor with GOES up to a maximum frequency of 1,200 Hz. (c) Core loss map of motor with NOES up to a maximum frequency of 1,200 Hz.

consideration has optimal grain size and texture as well as much smaller magnetic anisotropy in RD and TD as compared to GOES, which resulted in minimum eddy current and hysteresis losses. Furthermore, since the machine considered for analysis is a high–frequency, high–power motor, minimizing core losses in the electrical steel are the major contribution towards efficiency improvement [5]. The magnetic properties of NOES are highly dependent on the silicon content, cleanliness of the steel, the steel lamination thickness, grain size and crystallographic texture of the final laminations. Thus, for maximum magnetic flux density properties with minimally induced core losses, steel manufacturing process has to be optimized to obtain the required microstructure/texture with optimum performance characteristics [6].

5. Comparative Performance Analysis of High–Speed Motor: The Studied NOES vs. Commercial NOES

From the previous section, it can be seen that the motor with the studied NOES has better core loss characteristics and improved efficiency compared to the GOES counterpart. However, due to low magnetic flux density capability, the target torque performance of 70 Nm is not satisfied by the motor. In order to understand the inherent advantages of the studied NOES with optimized texture and grain size, the performance characteristics of the motor modeled in Section 2, is analyzed with the studied NOES and a commercially available NOES for the stator and rotor core material.

The torque– and power–speed characteristics obtained from the machines are illustrated in Figure 5(a). It can be seen that compared to the high–speed motor with commercial NOES, a 18.2% reduction in rated torque production, and a 14% reduction in output power capability are noticed for the motor with the studied NOES for a rated current of 78 A. However, due to the improved texture, the effective resistivity of the studied NOES is reduced, resulting in a 54.7% reduction in core losses when compared to the motor with commercial NOES. Furthermore, to match the torque rating, the current for the motor with the studied NOES material was increased from 78 A to 101 A. While this results in improved torque production, it directly reflects as increased motor copper losses and reduced efficiency. However, if the core loss improvement obtained from the studied NOES material is able to compensate for the increase in copper losses, the motor's efficiency with both the steel materials should be comparable. It can be seen from Figures 5(b) and (c) that, for almost the same torque rating, the studied NOES has 2.3 times lesser core losses and 1.6 times higher copper losses than the commercial NOES resulting in very close motor efficiencies. Thus, it can be concluded that with optimum grain size and texture, the NOES can be modified to obtain maximum magnetic properties with minimal core losses.

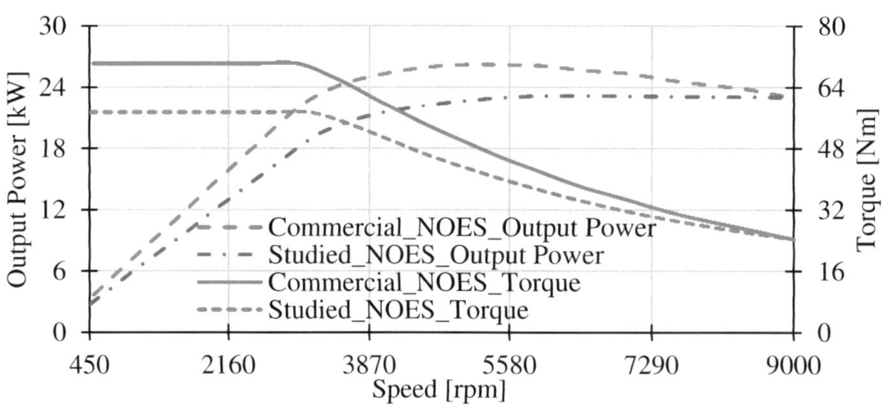

(a)

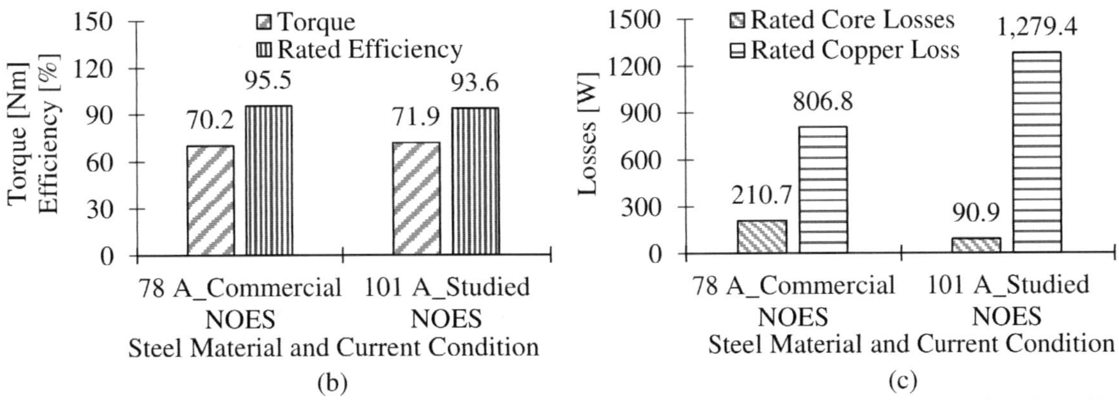

Figure 5. Comparison of high–speed motor performance characteristics with commercial and studied NOES materials. (a) Torque and power speed characteristics for a 78 A current rating. (b) Rated motor efficiency and torque production at 78 A for commercial NOES and 101 A for studied NOES. (c) Core and copper losses at 78 A for commercial NOES and 101 A for studied NOES.

6. Conclusions

This paper performs an exploratory performance analysis of NOES materials for high–speed traction motors emphasizing the importance of obtaining appropriate microstructure/texture with desired magnetic properties by controlling the annealing temperature and holding time. It was found that final annealing at 850°C for 60 minutes resulted in magnetically favorable θ–fibre texture and an optimum grain size, which give rise to better magnetic properties. A laboratory scale high–speed high–power traction motor was chosen for performance evaluation using finite element analysis. It was found that for the studied NOES, it can produce the same torque and power as a commercially available GOES, with significantly reduced core losses. Furthermore, when compared to a motor with a commercial NOES, the studied NOES also resulted in significantly reduced core losses and comparable rated efficiencies for the same torque production, highlighting the scope of improvement obtained by optimizing the microstructure/texture of the steel sheets.

7. References

[1] A. Saleem *et. al.*, "Effect of Shear Cutting on Microstructure and Magnetic Properties of Non–Oriented Electrical Steel," *IEEE Transactions on Magnetics*, vol. 52, no. 5, pp. 1–4, May 2016.

[2] S. Magdaleno–Adame *et. al.*, "Comparative Study of Grain Oriented and Non–Oriented Electrical Steels in Magnetic Shunts of Power Transformers," in proc. of *IEEE International Autumn Meeting on Power, Electronics and Computing* (ROPEC), Mexico, 2018, pp. 1–7.

[3] S. Takajo *et. al.*, "Effect of Silicon Content on Iron Loss and Magnetic Domain Structure of Grain–Oriented Electrical Steel Sheet," *IEEE Transactions on Magnetics*, vol. 54, no. 1, pp. 1–6, Jan. 2018, Art no. 2000206.

[4] Mehdi Mehdi, Youliang He, Erik Hilinski, Afsaneh Edrisy, "Non–oriented Electrical Steel with Core Losses Comparable to Grain–oriented Electrical Steel", Journal of Magnetism and Magnetic Materials, 2019, in press.

[5] R. Pei, L. Zeng and L. Gao, "Studies of Core Loss of Thin Non–Oriented Electrical Steel for Electrical Vehicle Traction Motors," in proc. of 21st *International Conference on Electrical Machines and Systems (ICEMS)*, Jeju, 2018, pp. 2715–2718.

[6] V. Manescu, G. Paltanea, H. Gavrila and I. Peter, "The influence of punching and laser cutting technologies on the magnetic properties of non–oriented silicon iron steels," in proc. of *International Symposium on Fundamentals of Electrical Engineering (ISFEE)*, Bucharest, 2014, pp. 1–4.

IOP Conf. Series: Materials Science and Engineering **654** (2019) 012005 doi:10.1088/1757-899X/654/1/012005

A Comparison Study of Magnetic Stirrer and Sonicator Technique to Disperse 1% Span20 Treated Layered Double Hydroxides (LDHs)

S Teanmetawong[1], T Chantaramanee[1], S Lhosupasirirat[1], A Wongariyakawee[4]and T Srikhirin[1, 2, 3]

[1] School of Materials Science and Innovation, Faculty of Science, Mahidol University, 999 Phutthamonthon Sai 4 Road, Salaya, Nakhon Pathom 73170, Thailand
[2] Physics Department, Faculty of Science, Mahidol University, Bangkok, 10400, Thailand
[3] NANOTEC Center of Excellence, Mahidol University, Bangkok, 10400, Thailand
[4] SCG Chemicals Co., Ltd., Siam Cement Road, Bang Sue, Bangkok, 10800, Thailand

toemsak.sri@mahidol.ac.th

Abstract. Effect of particle size reduction and dispersion of 2-dimensional material, layered double hydroxide (LDHs), prepared by co-precipitation method with the primary particle size of around 100 nm, was investigated. Two simple laboratory dispersing techniques including magnetic stirrer, and sonicator were applied to disperse agglomerated LDHs in 1 wt.% Span20 solution, act as a stabilizer. Magnetic stirrer showed a cluster of agglomerate particles with improved dispersion while, agglomerates with single stack LDHs were presence in case of sonicator product.

1. Introduction

Layered double hydroxides (LDHs) have a layered like structure as aluminosilicate or clay minerals. LDHs consist of a stack of positively metal hydroxide sheets. This charge is neutralized by interlayer anions solvated with water molecules in the interlayer. The chemical composition of LDHs structure can be written as $[M_{1-x}^{2+}\ M_x^{3+}\ (OH)_2]\ [A^{n-}]_{x/n}{\cdot}zH_2O$ where M^{2+} and M^{3+}refer to divalent and trivalent metal cations, respectively. A^{n-} represents anion [1], [2]. For a large-scale production, LDHs were synthesized by co-precipitation method which LDHs were generated by precipitation from mixture of metals salt and intercalated anions solution [3].

In order to disperse LDHs particle in organic solvent, surfactant is used for rending hydrophobicity to LDHs particles. To achieve homogenously fine colloidal dispersion, the agglomerate LDHs should be broken down and stabilized with proper type of surfactant. Thus, breaking down the large agglomeration without destroying the integrity of plate like particles and increasing interaction forces between LDHs are needed. There are several reported approaches which can be used to disperse nanoparticle. However, turbulence technique was reported to assist the adsorption of additives on fragments of the nanoparticles [4]. Magnetic stirrer provides a mechanical force through the rotating of permanent magnetic field to

Content from this work may be used under the terms of the Creative Commons Attribution 3.0 licence. Any further distribution of this work must maintain attribution to the author(s) and the title of the work, journal citation and DOI.
Published under licence by IOP Publishing Ltd

induce the movement of the magnetic bar that immersed in the suspension. Sonicator uses an ultrasound irradiation to agitate the fluid, water, normally, by bubbled cavitation effect [5]. These bubbles collapsing creates energy with combined high pressure and temperature.

In this study, the reduction of agglomerate LDHs to LDHs stacking, individual plate like structure of LDHs, using magnetic stirrer and sonicator were investigated. Span20, an industrial non-ionic surfactant type with the chemical formulation of $C_{18}H_{34}O_6$, was used as the stabilizer, figure 1a. Span20 possesses hydrophile-lipophile balance (HLB) value of 8.6 was used as stabilizer to prevent re-agglomeration [4].

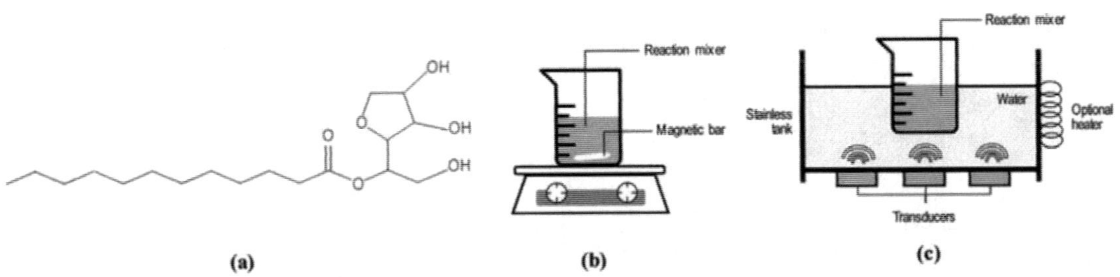

Figure 1. (a) molecular structure of Span20, (b) magnetic stirrer and (c) sonicator model

2. Materials and methods

2.1 Materials

Unmodified surface LDHs wet cake with primary particle size, approximately 100 – 200 nm, was used as raw material and was kindly supplied by SCG Chemicals Co., Ltd., Thailand. Span20 was purchased from LobaChemie Pvt., Ltd., Mumbai, India. Methyl Ethyl Ketone or Butanone (MEK) was purchased from RCL Labscan.

Reversed Osmosis (RO) water was produced by 50 GPD auto flash (Unipure, imported by Function International Co., Ltd., Bangkok, Thailand). Magnetic stirrer was carried out by using laboratory hotplate stirrer (HTS-1003, LMS, Tokyo, Japan). Sonicator (S 70 H, Elmasonic, Singen, Germany) with fixed power at 70 W was used.

2.2 Sample preparation and dispersion

1 wt.% Span20 was mixed with 200 mL RO water at room temperature. Suspension of 5 wt.% LDHs wet cake was added in Span 20 solution. The mixture was stirred for 2 hours at room temperature. For sonicator dispersion, the suspension was mixed, stirred 1 hour and sonicated for 1 hour. The sample obtained from both techniques was separated by using centrifuge and dried in an oven at 80 °C, yielding treated LDHs powder.

2.3 Characterization

The dispersion test was carried out by redispersion of treated LDHs powder in MEKwith 0.0625 wt.% and placed in sonicator for 30 minutes. The particle morphology and its distribution were observed by field emission scanning electron microscope (FE-SEM SU8010, HITACHI, Japan). Ultraviolet-visible spectroscopy (Cary 60 UV-vis, Agilent Technologies, USA) was used for monitoring the transmission of the suspension. Particle size distribution was measured by Zetasizer Nano ZS (Malvern, UK) with the measurement angle of 173° backscatter. Fourier-transform infrared spectroscopy (FTIRNICOLET 6700, Thermoscientific, USA) with ATR mode was used for identifying the existence of Span20 on LDHs. The scanning range was from 4000 – 650 cm^{-1} wavenumber with %transmission as Y-axis. Number and resolution of scan was set as 6.4 and 2, respectively.

3. Result and discussion

3.1 Raw LDHs characteristic

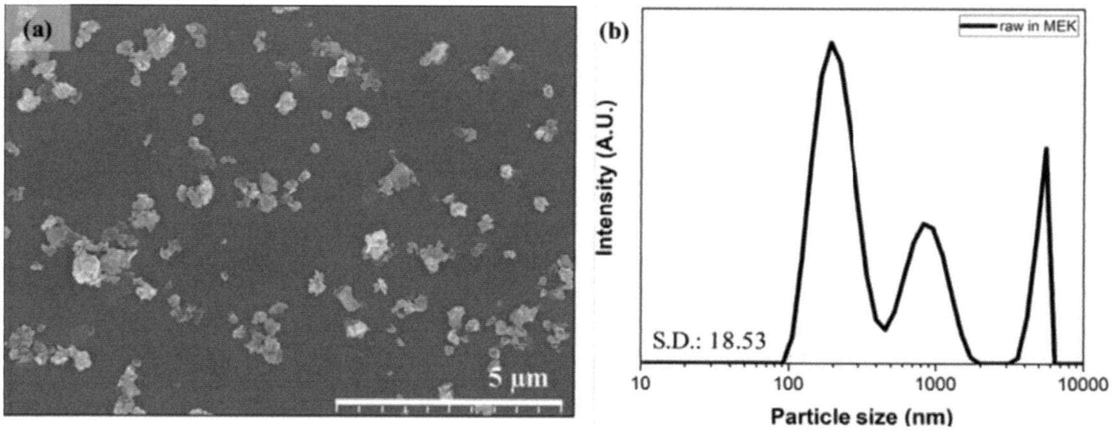

Figure 2. (a) FE-SEM image and (b) particle size distribution curve of raw LDHs dispersed in MEK

Figure 2a, showed a FE-SEM image of as received LDHs. Agglomerate LDHs were found with dense packing and poor dispersion in MEK. Particle size is ranging from a few micrometers down to 100 – 200 nm. They were the agglomerate of the primary particle size. Particle size distribution (PSD), figure 2b, was a tri-modal broad distribution ranging from a few hundred nanometer upto micrometer range.

3.2 Comparison of two dispersing techniques

Figure 3a showed the appearance of 0.0625 wt.%LDHs suspension of as received LDHs and Span20 treated LDHs by stirrer and sonicator. The background cannot be seen in raw LDHs sample where the lowest %transmittance was obtained from UV-vis spectra due to the scattering of the agglomerate LDHs, figure 3b. For stirred and sonicated sample, a higher %transmittance can be observed with a clearer background. Unexpectedly, higher %transmittance was found in stirred sample. The transmission is a direct variation to the particle size distribution. The lower transmission in all wavelengths indicated a lower level of transmission of the light due to the light scattering of the suspended particle. The level of transmission was quantified in figure 3b.

Figure 4 showed FE-SEM images of 1 wt.% Span20 treated LDHs prepared by using stirrer and sonicator. Particle size reduction and Span20 treatment promoted a better dispersion and size reduction when compare to raw LDHs. It can be seen that, there were still large agglomerate particles or cluster around submicron range in stirred sample, figure 4a. Magnetic stirrer cannot completely disintegrate large agglomerate particles into the smaller particle. Individual LDHs stacking could not be obtained by stirrer. For sonicated sample, large agglomerate particles and smaller particles were found in all the area, figure 4b.

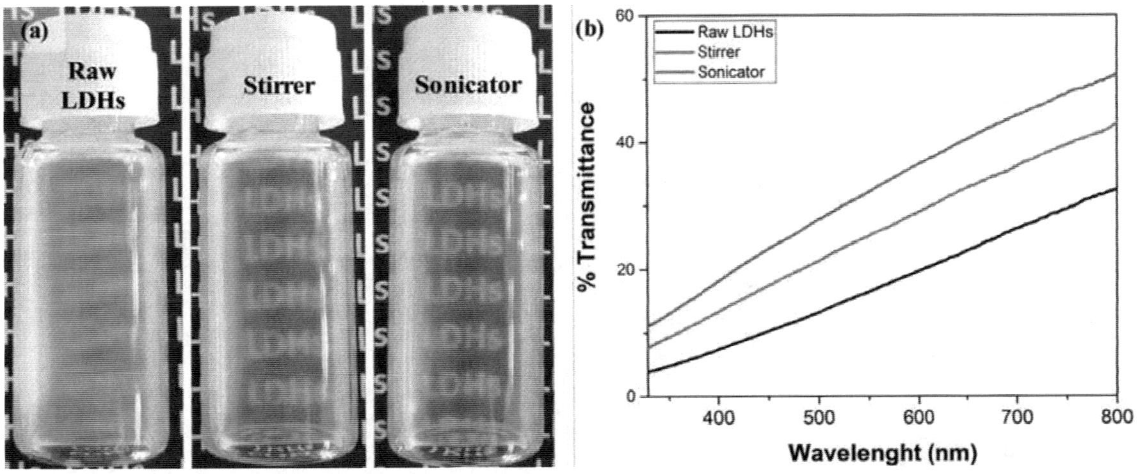

Figure 3. (a) appearance of 0.0625 wt.% of raw LDHs, magnetic stirrerand sonicatorfor1 hourof 1% Span20 treated LDHs disperse in MEK, respectively and (b) UV-vis spectra with relation of %transmittance and wavelength from 329 to 800 nm of each sample

This correspond to the particle size distribution from DLS, figure 4c. Agglomerate was still found in all samples, before and after dispersing process. Multimodal distribution peak scan was found in stirred sample, which meant there were variation of particle size. Agglomerate particles in stirred sample were found in similar size to the prior dispersing process. The agglomerate size of sonicated sample was reduced where both fine and coarse particles were observed. This might be attributed to the sonicator can only scour the surface particles from the agglomerate particles [6], resulting in individual LDHs stacks with agglomerates. However, agglomerates cannot be broken further because long sonication time may induce re-agglomeration [6].

From FTIR spectra, figure 5, there were a broadband from O-H vibration, water molecule bending and a sharp band from intercalated carbonate molecules appeared around 3420, 1620 and1364 cm^{-1} wavenumber in all samples. The characteristic bands of LDHs were also observed at 430 and 560 cm^{-1} which attributed to the lattice vibration of M-O and M-O-M, respectively [7]. After dispersion, it can be confirmed that Span20 appeared on LDHs particle from both techniques. Seeing that, there are CH stretching, C=O and C-O-C which are characteristic peaks of Span20, located approximately at 2850 – 3000, 1650 – 1800 and 1180 – 1200 cm^{-1} wavenumber [8], which cannot be seen in raw LDHs. Hence, the result indicated that, stirrer and sonicator can be used to treat Span20 on LDHs particles.

4. Conclusion

Laboratory available technique, magnetic stirrer and sonicator, potentially stabilized LDHs particle by Span20. The improvment in dispersion can be obtained after dispersing process for an hour. Sonicator can separate the large agglomerate into single stack LDHs surrounding with the agglomeration. For stirrer, the agglomeration cannot be completely broken. Because, the input energy provided by magnetic stirrer and sonicator was not sufficient enough to break down all the large agglomerate particles into individual particles.

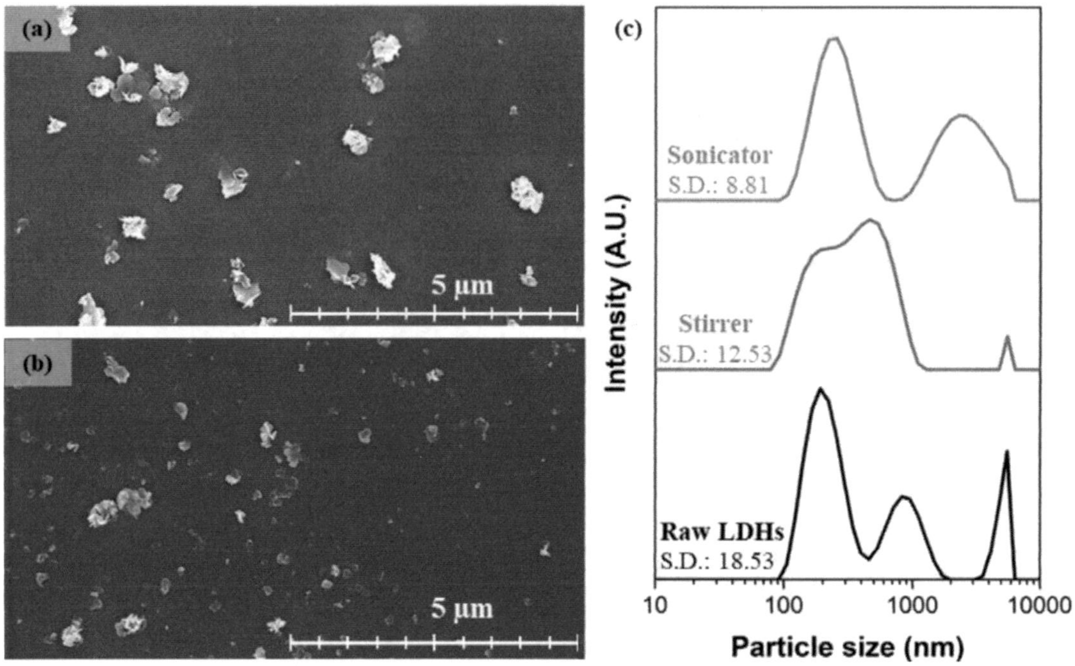

Figure 4. FESEM images of (a) stirred, (b)sonicated sample and (c) particle size distribution of (black) raw LDHs, (blue) magnetic stirrer and (green) sonicator for 1 hour

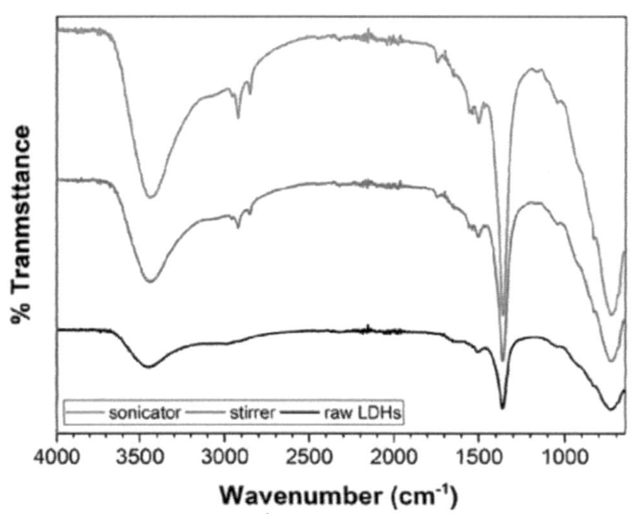

Figure 5. FTIR spectra from 4000 to 650 cm^{-1} wavenumber of each dispersion technique (black) raw LDHs, (blue) magnetic stirrer and (green) sonicator for 1 hour

5. References

[1] Wang Q and O'Hare D 2012 Recent Advances in the Synthesis and Application of Layered Double Hydroxide (LDH) Nanosheets *Chem. Rev.* **112** 4124–55

[2] Benício L P F, Silva R A, Lopes J A, Eulálio D, dos Santos R M M, De Aquino L A, Vergütz L, Novais R F, Da Costa L M, Pinto F G and Tronto J 2015 Layered double hydroxides: Nanomaterials for applications in agriculture | Hidróxidos duplos lamelares: Nanomateriais para aplicações na agricultura *Rev. Bras. Cienc. do Solo***39** 1–13
[3] Bravo-Suárez J J, Páez-Mozo E A and Oyama S T 2004 Review of the synthesis of layered double hydroxides: A thermodynamic approach *Quim. Nova***27** 601–4
[4] Müller F, Peukert W, Polke R and Stenger F 2004 Dispersing nanoparticles in liquids *Int. J. Miner. Process.***74** 31–41
[5] Pal N, Kumar N and Mandal A 2019 Stabilization of Dispersed Oil Droplets in Nanoemulsions by Synergistic Effects of the Gemini Surfactant, PHPA Polymer, and Silica Nanoparticle *Langmuir***35** 2655–67
[6] Sumitomo S, Koizumi H, Uddin M A and Kato Y 2018 Comparison of dispersion behavior of agglomerated particles in liquid between ultrasonic irradiation and mechanical stirring *Ultrason. Sonochem.***40** 822–31
[7] Tong M, Chen H, Yang Z and Wen R 2011 The effect of Zn-Al-hydrotalcites composited with calcium stearate and β-diketone on the thermal stability of PVC *Int. J. Mol. Sci.***12** 1756–66
[8] Cortés F B, Lozano M, Santamaria O, Marquez S B, Zapata K, Ospina N and Franco C A 2018 Development and evaluation of surfactant nanocapsules for chemical Enhanced Oil Recovery (EOR) applications *Molecules***23** 1–19

IOP Conf. Series: Materials Science and Engineering **654** (2019) 012006 doi:10.1088/1757-899X/654/1/012006

Effect of Reacting Temperature on the Microstructure and Electronic Property of Self-ordered TiO_2 Nanotube Arrays

Jiangyi Yan[1]

[1] Department of mechanical engineering, Tsinghua University, Beijing, 100084, China

2115731095@qq.com

Abstract. At present, with the rapid development of nanoscience, nanotubes have become an important part of nanomaterials. As a new type of multifunctional semiconductor nanotube material, the self-ordered TiO_2 nanotube arrays (TNTs) have excellent optical and electrical properties, and they have a very wide application prospects in modern industrial fields. By using the method of anodic oxidation, the specific influence of reaction temperature on the microstructure and electrochemical properties of TNTs under the condition of acid system was investigated by the way of controlling variables. The results show that the lengths of TNTs and the thickness of the barrier film decrease with increasing temperature, and which is opposite in the radius and tube wall thickness. Additionally, in electrochemical properties of TNTs, the equivalent electrochemical impedance and crystallinity usually decrease with increasing of the temperature.

1. Introduction

In many semiconductor nanomaterials, TNTs are very important multifunctional inorganic nanomaterials [1], they have extremely stable physical and chemical properties, and compared to TiO_2 nanoparticles [2], TNTs have high surface area, unique optical and electrical properties and they also have geometric features [17-19], which are more beneficial to the electronic transmission. Therefore, TNTs have a very wide range of applications in the field of modern industry [3]. First, because of its high photoelectric conversion efficiency [16], TNTs are used as an anode material for dye sensitized solar cells [19]. Second, TNTs are ideal materials for the photodissociation of hydrogen [8-10]. Third, as titanium has good biocompatibility and degeneracy, TNTs are also widely used as a raw material for some biological scaffolds [11-14]. In addition, by means of cyclic voltammetry and constant current measurement, some researchers studied the sensitivity of TiO_2 nanotube arrays with adsorption of horseradish peroxidase (HRP) and thionine chloride (TH) for hydrogen peroxide [4-7].

A detailed investigation on the microstructure and electrochemical properties of TNTs is beneficial to the application, so it is necessary to study the influences of fabrication temperature on the microstructure and electronic property of TNTs. Although a great deal of researchers paid attention to this issue [15],

Content from this work may be used under the terms of the Creative Commons Attribution 3.0 licence. Any further distribution of this work must maintain attribution to the author(s) and the title of the work, journal citation and DOI.
Published under licence by IOP Publishing Ltd

there is no systematic researches and conclusion about the temperature influences on the formation parameters on the microstructure and characterization of TNTs, the purpose of this paper is mainly to investigate the concrete and comprehensive influences about the temperature on the microstructure and electronic property of TNTs by SEM, Roman spectrum, EIS and Mott-Schottky plot.

2. Experimental

2.1 Preparation of TiO_2 nanotube arrays

At present, the methods of preparing titanium dioxide nanotube arrays by anodic oxidation are mainly divided into two kinds: the acid system and the alcohol system: electrolyte solution is fluorine-containing inorganic acidic aqueous solution and electrolyte solution is fluorine-containing organic solution [3-5]. In this study, titanium nanotube arrays were fabricated in the acid system. The basic process of the experiment was as follows: Titanium sheets were deposited in the mixed solution of 0.15mol/LHF+1mol/LH2SO4 in electrolyte solution for 2 hours at the experimental temperature of 0, 30, 60 and 90°C respectively, and the I-T curves of the film forming process were recorded. MS and EIS tests were carried out immediately in 0.5mol/L Na2SO4 solution, and then sintered and sintered. After completion, the SEM is observed on the side and the Raman spectrum is finally detected. For the temperature change and maintenance, DF-II collector magnetic heating mixer and DC-4015 energy-saving intelligent thermostat are needed.

2.2 Electrochemical measurement and microstructure observation

Electrochemical studies were conducted using the Solatron 1260+1287 Electrochemical Workstation. A conventional three-electrode glass cell was used to carry out the electrochemical studies, in which, Pt foil was the counter electrode and saturated calomel electrode (SCE) was used as the reference electrode.

EIS measurement was measured in the frequency range of 5 mHz to 100 kHz with the applied alternating sinusoidal potential of 10 mV. After each experiment, the impedance data were displayed as Nyquist and Bode in phase format plots.

Mott-Schottky plot was carried out in the potential range of -1.0V to 1.0V with a scanning rate of 20mV/S. In order to eliminate the frequency dispersion, the frequency of 1KHz was used as the applied frequency.

Micro-Raman spectra were obtained using the Horiba Jobin HR800 Raman spectrometer with 15mW of a 514 nm laser.

The surface and cross section morphologies of the TiO_2 NTs were examined by SUPRA55 Field Emission Scanning Electron Microscope.

3. Results and discussions

3.1 Effect on the micromorphology of TiO_2 nanotube arrays

TNTs fabricated at different temperatures were enlarged 100 thousand times under an electron microscope for observation, and the micro-morphologies of the side and front were obtained in SEM images as shown in figure 1 and figure 2 and the characteristic size of TNTs at different temperatures is calculated as Table 1 shows. It is obviously seen that the length of nanotubes and the thickness of barrier layer decrease with the increasing temperature and the rate of change decreases. And, the diameter and wall thickness of nanotubes increase with the increasing temperature, and the rate of change decreases too. In addition, it can be seen that the change rate of the characteristic sizes of nanotubes with temperature changes from large to small is length > barrier thickness > diameter > wall thickness, that is to say, temperature has the greatest influence on the length and the smallest influence on wall thickness.

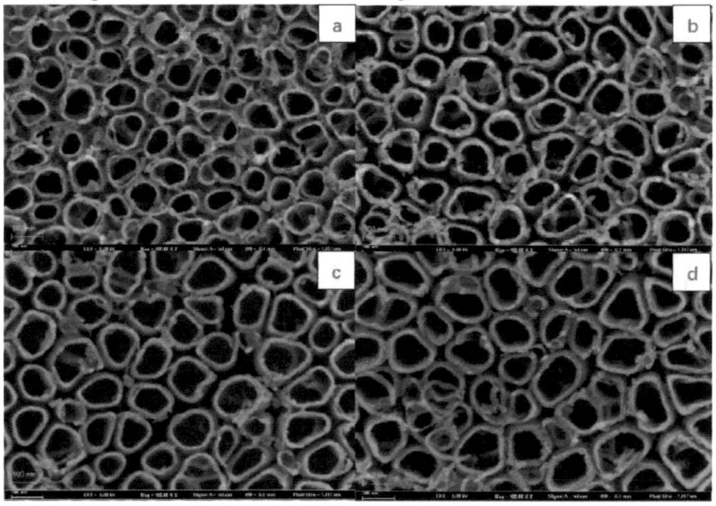

Figure 1. The side-section SEM images of TiO_2 nanotube arrays fabricated by anodization at different temperatures for 2h in 0.15M HF+1.0M H_2SO_4 solution, (a)0°C, (b)30°C, (c)60°C, (d)90°C.

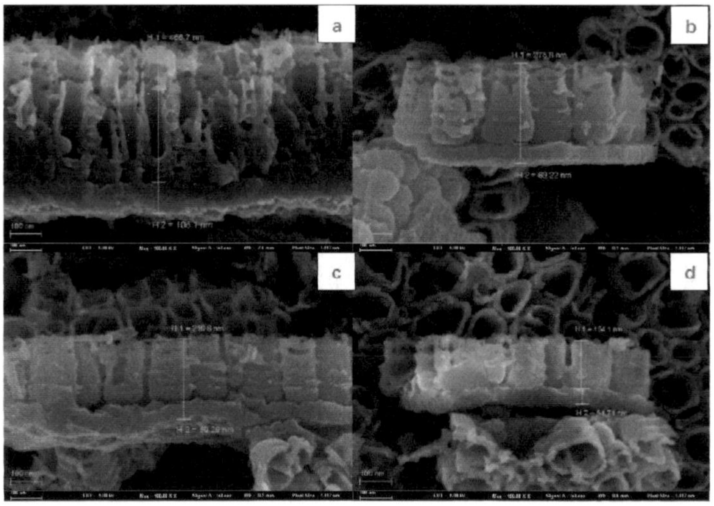

Figure 2. The front-section SEM images of TiO_2 nanotube arrays fabricated by anodization at different temperatures for 2h in 0.15M HF+1.0M H_2SO_4 solution, (a)0°C, (b)30°C, (c)60°C, (d)90°C.

Table 1. the characteristic size of TNTs fabricated at various temperatures for 2h in 0.15M HF+1.0M H_2SO_4 solution.

Temperatures/°C	Diameter/nm	Wall thickness /nm	Length/nm	Barrier thickness /nm
0	61.41	13.40	466.7	106.1
30	79.27	14.89	275.8	69.22

IOP Conf. Series: Materials Science and Engineering **654** (2019) 012006 doi:10.1088/1757-899X/654/1/012006

60	94.90	15.63	216.6	60.29
90	101.6	16.02	164.1	54.71

3.2 Effect on the electronic property of TiO_2 nanotube arrays

In order to detect the influence of temperature on the electronic properties of TNTs, the EIS and MS curves of TNTs are measured in 0.5M Na2SO4 solution. The EIS and Mott-Schottky curves of TNTs obtained at different temperatures are shown in figure 3. The EIS curve shows that the radius of different curves decreases with the increased temperature, which indicates equivalent electrochemical impedance decreases continuously with the increased temperature. And, from the MS curve, reversed sigmoidal plots can be observed with an overall shape consistent with that typical for n-type semiconductors, and the slope of the linear region sharply decreases with the increased temperature, that is to say, the flat-band voltage of TNTs is increasing, which means conductivity of TNTs is getting stronger and stronger according to the carrier density formula of semiconductors. Also, it indicates the increased donor density for the TNTs electrode when the temperature is increasing, and a higher donor density is an indicator of the enhanced pitting corrosion susceptibility.

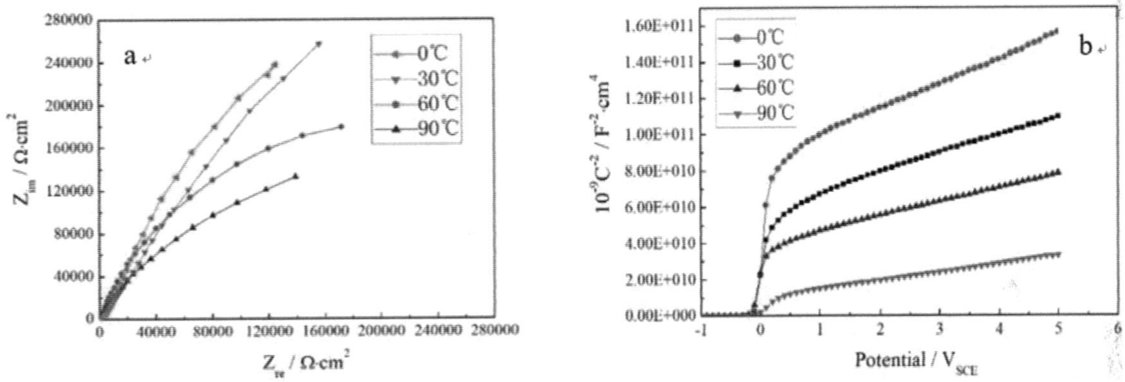

Figure 3. EIS (a) and MS (b) curves of TNTs at different temperatures for 2h in 0.15M HF+1.0M H_2SO_4 solution.

In addition, the phase composition of TNTs is tested by micro-Raman spectroscopy utilizing variable excitation wavelengths, and Figure 4 shows four micro-Raman spectra of TNTs fabricated at different temperatures. Four Raman bands are clearly seen at 143, 393, 512, and 632 cm^{-1}, respectively [20]. And, the characteristic peaks of Raman spectra of the TNTs prepared at different temperatures are basically the same, and there is no change, which indicates that temperature has little effect on the microstructure of the TNTs.

In addition, with the increase of temperature, the overall intensity of Raman spectroscopy is decreasing, which shows that the temperature is negatively related to the crystallinity of TNTs. In other words, it means that the crystallinity of TNTs decreases with the increased temperature.

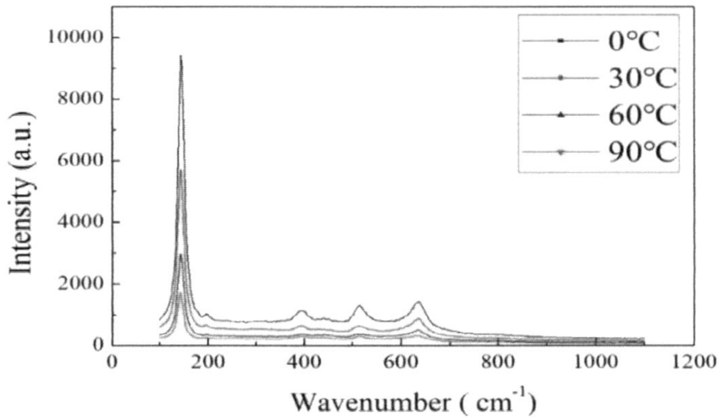

Figure 4. Raman spectrum of TNTs fabricated at different potentials for 2h in 0.15M HF+1.0M H_2SO_4.

4. Conclusion

Temperature has a great influence on the microstructure and electronic properties of TNTs fabricated by anodic oxidation in the acid environment, conclusions can be drawn as following:

1) the length and the thickness of barrier layer of TNTs decrease with the increasing temperature and the diameter and wall thickness increase with the increasing temperature. And, it can be seen that the change rates of four characteristic sizes are length > barrier thickness > diameter > wall thickness with temperature changing from large to small;

2) the stability of TNTs is decreasing and the conductivity is increasing with the increase of temperature.

3) the crystallinity of TNTs is decreasing with the increase of the temperature.

Acknowledgment

1. This work is supported by the State Key Laboratory of Tribology in Tsinghua university of China.
2. This work is supported by the National Science and Technology Major Project 2017ZX04010001 of China.
3. This work is supported by Research Room of Intelligent CNC System of Department of mechanical engineering in Tsinghua university of China.

5.References

[1] Wu P and Xu Z 2005 Silanation of nanostructured mesoporous magnetic particles for heavy metal recovery *Industry and Engineering Chemistry Research*. 44(2005)816-825.

[2] Hiroaki T, Jan M, Andrei G, Arlindo S, Luciano T and Patrik S 2007 Characterization of electronic properties of TiO_2 nanotube films *Corrosion Science*. 49 (2007) 203–210.

[3] Fujishima A and Honda K 1972 Electrochemical photolysis of water at a semiconductor electrode *Nature*. 238(1972)37-39.

[4] Shankar K, Mor G K, Prakasam H E, Yoriya S, Paulose M, Varghese O K and Grimes C A 2007 Highly-ordered TiO_2 nanotube arrays up to 220μm in length: use in water photo electrolysis and dye-sensitized solar cells *Nanotechnology* 8(2007)065707.

[5] Varghese O K, Gong D, Paulose M and Grimes 2003 Extreme changes in the electrical resistance of

titania nanotubes with hydrogen exposure *Adv. Mater.* 15(2003)624-627.

[6] Guan D S, Paul J, Hymel and Wang Y. Growth 2012 mechanism and morphology control of double-layer and bamboo-type TiO_2 nanotube arrays by anodic oxidation *Electrochemical Acta.* 83(2012) 420-429.

[7] Regonini A, Schmidt C Aneziris T and Graule F J 2015 Clemens Impact of the Anodizing Potential on the Electron Transport Properties of Nb-doped TiO_2 Nanotubes *Electrochemical Acta.* 169(2015) 210-218.

[8] Paulose M, Shankar K, Varghese O K, Mor G K and Grimes C A 2006 Application of highly-ordered TiO_2 nanotube-arrays in heterojunction dye-sensitized solar cells *Corros. Sci* 39(2006)2498.

[9] Lu X H, Wang G M, Zhai T, Yu M H, Gan J Y, Tong Y X and Li Y 2012 Hydrogenated TiO_2 Nanotube Arrays for Supercapacitors *Nano Lett.* 12(2012)1690.

[10] Ohtsuka T, M. Masuda M and Sato N 1985 Ellipsometric Study of Anodic Oxide Films on Titanium in Hydrochloric Acid, Sulfuric Acid, and Phosphate Solution. 132(1985)787.

[11] Lu H F and Li Feng 2008 Amorphous TiO_2 nanotube arrays for low-temperature oxygen sensors Nanotechnology 19 405504.

[12] Nelson J C and Oriani R A 1993 Stress generation during anodic oxidation of titanium and aluminum. 34(1993)307.

[13] Regan B O, Moser J, Anderson M, and Graetzel M 1990 Vectorial electron injection into transparent semiconductor membranes and electric field effects on the dynamics of light-induced charge separation *J. Phys. Chem.* 84(1990)8720.

[14] Leach J S L and Pearson B R 1988 Crystallization in anodic oxide films *Corros. Sci.* 28(1988)43.

[15] Armstrong N R and Quinn R K 1977 Auger and X-ray photoelectron spectroscopic and electrochemical characterization of titanium thin film electrodes *Surf. Sci.* 67(1977)451.

[16] Shibata T and Zhu Y C 1995 The effect of film formation conditions on the structure and composition of anodic oxide films on titanium *Corros. Sci.* 37(1995)253.

[17]Krishnamurthy B, White R E and Ploehn H J 2005 Electric field strength effects on time-dependent passivation of metal surfaces *Electrochemical Acta.* 47(2002)2505.

[18] Prakasam H E, Paulose M and Varghese O K 2007 TiO_2 nanotube arrays of 1000μm length by anodic oxidation of titanium foil *Journal of Physical Chemistry C.* 111:14993-14996.

[19] Schmidt A M, Azambuja D S and Martini E M A 2006 Semiconductive properties of titanium anodic oxide films in McIlvaine buffer solution *Corros. Sci.* 48(2006)2901.

[20] Paulose M, Shankar K and Yoriya S 2006 Anodic growth of highly-ordered TiO_2 nanotube arrays up to 134 μm in length *Journal of Physical Chemistry B.* 110 (2006): 16179-16184.

IOP Conf. Series: Materials Science and Engineering **654** (2019) 012007 doi:10.1088/1757-899X/654/1/012007

Investigation of Citrate Modified Layered Double Hydroxides by Bead Milling Technique

T Chantaramanee[1], S Teanmetawong[1], S Lhosupasirirat[1] A Wongariyakawee[4] and T Srikhirin[1, 2,3]

[1] School of Materials Science and Innovation, Faculty of Science, Mahidol University, 999 Phutthamonthon Sai 4 Road, Salaya, Nakhon Pathom 73170, Thailand
[2] Department of Physics, Faculty of Science, Mahidol University, Bangkok, 10400, Thailand
[3] NANOTEC Center of Excellence, Mahidol University, Bangkok, 10400, Thailand
[4] SCG Chemicals Co., Ltd., Siam Cement Road, Bang Sue, Bangkok, 10800, Thailand

toemsak.sri@mahidol.ac.th

Abstract. Agglomeration and low stability of layered double hydroxides (LDHs) suspended in water significantly affects its properties for water base applications. In this study, the stability of LDHs suspended in water was improved by using bead milling for size reduction and by surface treatment with tri-sodium citrate. The result showed that bead milling technique can reduce the agglomeration and induce the adsorption of tri-sodium citrate on LDHs without intercalation. Following redispersion in water, the stability of citrate-modified LDHs was >1 hr with particle size around 177 nm, whereas the particle size of unmodified LDHs was around 40 μm with fast sedimentation.

1. Introduction

Layered Double Hydroxides (LDHs) are classified as a 2 dimensional (2D) nanomaterials containing positive charged brucite-like layers which are neutralized by anions, often hydrated with water, in the interlayers. The general chemical formulation for LDHs is $[M^{II}{}_{1-x}M^{III}{}_{x}(OH)_2]^{X+}\,(A^{n-})_{x/n}\circ yH_2O$ where M^{II}, M^{III} are divalent and trivalent cation, respectively [1]. LDHs possess several unique properties for example flame retardance, high mechanical properties, and anion exchangeability. They have been used in several applications, including bio nanotechnology [3], and as polymer additives [2]. LDHs are generally prepared by co-precipitation. In their agglomerated form, the individual particles (LDHs stack) are held together by Vander Waal interaction [4]. LDHs sheets have high surface energy due to abundant hydroxyl groups leading to low dispersibility in water and organic solvents. Since environmental problem can be caused by harmful organic solvents, water based formulations are gaining popularity.

Surface modification is a traditional method that is used to improve dispersibility of nanoparticles in liquid media. An anionic surfactant was considered in this work because the net cationic charge on LDHs layers may induce an electrostatic attraction between the positively charged LDHs and negatively charged anions. Tri-sodium citrate (citrate) is a type of carboxylic acid salt that is widely used to promote the suspension of nanoparticle in aqueous phase [5,6]. Citrate is a harmless and inexpensive material with chemical structure as shown in Figure 1.

Content from this work may be used under the terms of the Creative Commons Attribution 3.0 licence. Any further distribution of this work must maintain attribution to the author(s) and the title of the work, journal citation and DOI.
Published under licence by IOP Publishing Ltd

In this work, the effect of mechanical forces on the LDHs particle size was investigated. Bead milling is the technique that gives highest grinding energy, compared to other dispersion techniques [7]. Impact and shear forces are produced by collision between the grinding beads and particles.

The objectives of this work were to reduce the agglomerated particle size by using bead milling process and to improve the dispersibility of LDHs in water by using tri-sodium citrate as a modifier. The existence of citrate on LDHs was confirmed by Fourier Transform Infrared Spectroscopy (FT-IR). Whereas, the crystal structure of citrate-modified LDHs was observed by X-ray diffraction technique (XRD). The particle size and dispersibility of citrate-modified LDHs in water was also investigated.

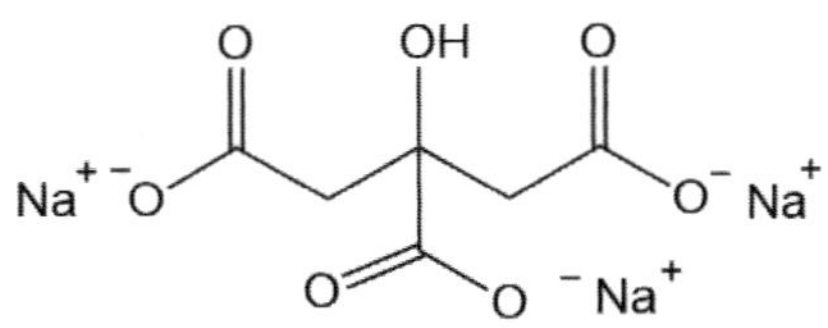

Figure 1. Chemical structure of tri -sodium citrate

2. Experimental

2.1 Materials

LDHs nanoparticle in a wet cake form with a primary particle size of 50-200 nm in the lateral dimension and around 27 nm for the stack was obtained from SCG Chemicals Co., Ltd., (Bangkok, Thailand). Analytical grade Tri-sodium Citrate was purchased from Ajax Finechem Pty Limited (Australia). Reverse osmosis water (RO) from Unipure RO-50 GPD imported by Function International Co., Ltd., (Bangkok, Thailand) was used for all experiments.

2.2 Citrate-modified LDHs preparation

Citrate-modified LDHs was prepared by mixing LDHs with 200 mL of tri-sodium citrate in RO water. The suspension was stirred by a magnetic stirrer (HTS-1003, LMS CO., LTD., Tokyo, Japan) before milling (Minicer, NETZSCH GmbH & Co. Holding K G, Germany) for various durations (15-60 minutes). The sample was cleared of excess citrate by centrifugation and dried overnight in oven (572F WTB BINDER, BINDER GmbH, Germany) at 80°C.

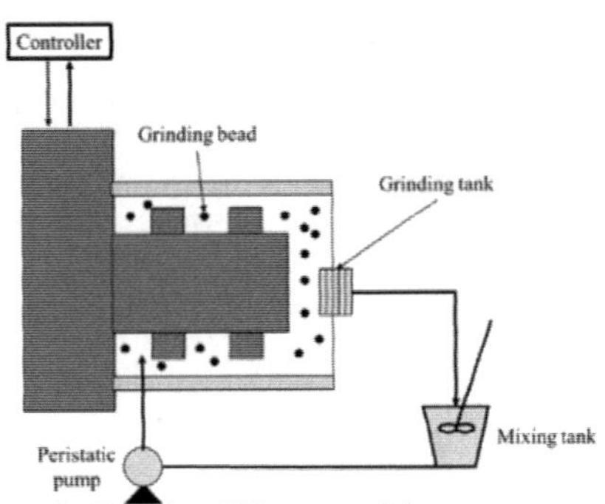

Figure 2. Bead milling machine

2.3 Characterizations

The diffraction pattern of LDHs and citrate-modified LDHs were observed by powdered X-ray diffraction (XRD) technique, (D8 Advance, Bruker, Germany), with Cu Kα radiation source ($\lambda = 0.154$ nm), step size 0.04° and time step 1.00 s.

The LDHs and citrate-modified LDHs were investigated by FTIR (NICOLET 6700, thermo scientific, USA) in ATR mode to confirm the existence of surfactants on LDHs particle.

The LDHs and citrate-modified LDHs were dispersed in RO water by a table top sonicator (Elmasonic S 70 H, Singen, Germany) for 30 minutes. The average particle size and zeta potential of both suspensions were measured by Zetasizer Nano ZS (Malvern, UK) with a measurement angle of 173° backscatter. The measurements were repeated thrice to calculate the average particle size.

The particle morphology and particle distribution of LDHs were observed by field emission scanning electron microscope (FE-SEM) (HITACHI, SU8010, Tokyo, Japan). Before measurement, the sample suspension was dropped on a clean silicon wafer and dried in an oven.

3. Result and discussion

Bead milling technique was used for dispersing and modifying tri-sodium citrate on LDHs particle. Grinding parameters can be varied including milling time, shaft speed, feed rate, and bead size [8]. To observe the optimum milling time, the citrate-modified LDHs were redispersed in water. The particle size of citrate-modified LDHs redispersed in water at each milling time were as shown in Figure 3. At 0 minutes the average particle size was 519 nm. After 15 and 30 minutes of milling, the average particle size was reduced to around 176 nm. After 40 minutes of milling time, the average particle size increased to 268 nm which implied as the re-agglomeration of LDHs particle. At 60 minutes the average particle size decreased dramatically to 50 nm. It implied that longer milling time, high energy, may cause breaking of primary particle.

Zeta potential of citrate-modified LDHs after redispersion in water was shown in Figure 3a. The zeta potential showed negative sign with magnitude higher than 30 mV, which meant that the LDHs particles had high stability in water after modification with tri-sodium citrate [9]. A general trend of decreasing zeta potential with increasing milling time was noted.

The results showed that bead milling can reduce LDHs particle size and modify citrate on LDHs. The optimum milling time for dispersion and modification of LDHs by citrate in water was 15 minutes because it was the shortest time that provided LDHs in primary size and zeta potential lower than -30 mV. This optimum milling time was used for the preparation of citrate-modified LDHs.

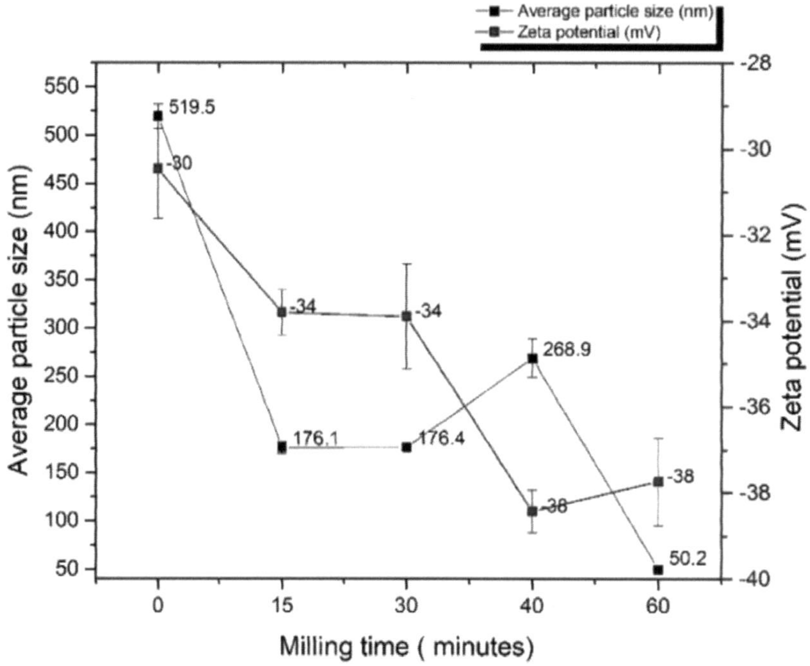

Figure 3. Average particle size of citrate-modified LDHs redispersion in water at each milling time

FTIR results of citrate-modified LDHs particle after 15 min of grinding was shown in Figure 4a. These results were used to confirm the existence of citrate on LDHs particle after modification. For raw LDHs, the adsorption band at 1362 cm^{-1} revealed anti-symmetric stretching mode of carbonate ion

IOP Conf. Series: Materials Science and Engineering **654** (2019) 012007 doi:10.1088/1757-899X/654/1/012007

that intercalated into the interlayered space [10]. Characteristic bands of tri-sodium citrate were seen at 1417 cm^{1} and 1586 cm^{-1}, which were assigned to symmetric of -COO^{-} and asymmetric stretching [11]. The FTIR result of citrate-modified LDHs revealed the -COO^{-} stretching of tri-sodium citrate. First band was small shoulder at 1568 cm^{-1} and second band was at 1362 cm^{-1}; however, it combined with peak of carbonated intercalated ion of LDHs. The FTIR result confirmed that tri-sodium citrate existed on LDHs particle after bead milling process. However, the intensity of characteristic bands of tri-sodium citrate on citrate-modified LDHs was low when compared to LDHs because of the small amount of tri-sodium citrate present on citrate-modified LDHs. Other techniques, such as total organic carbon analysis, could be useful to further verify the existence of citrate on LDHs particles.

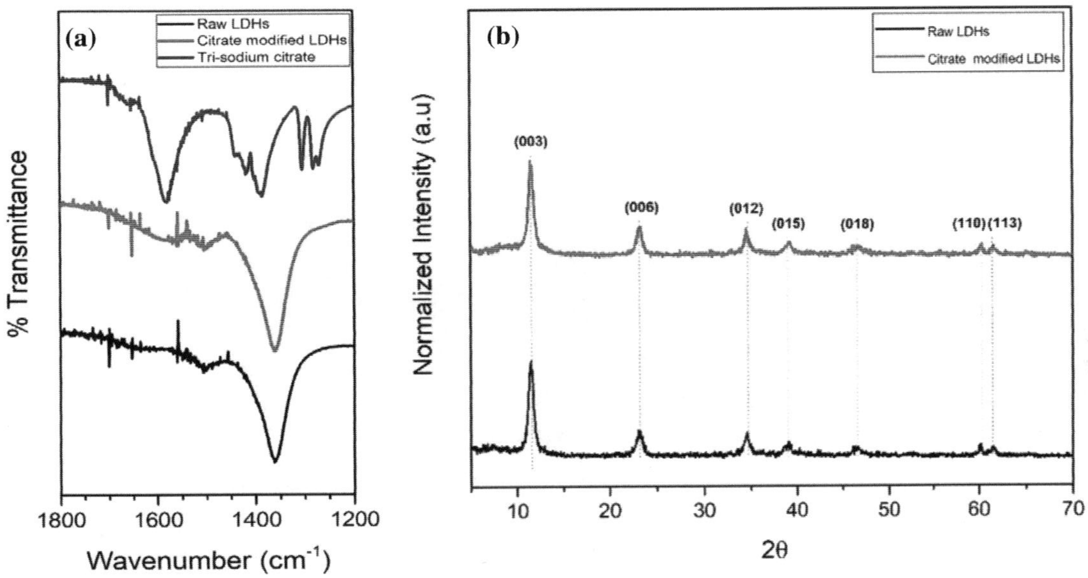

Figure 4. (a) FTIR result and (b) XRD result of raw LDHs and citrate modification LDHs by bead milling 15 minutes

The XRD pattern of Raw LDHs and citrate-modified LDHs after 15 minutes grinding was shown in Figure 4b. The 003 peaks of Raw LDHs and citrate-modified LDHs were located at the same position at $2\theta = 11.52$ °. The basal spacing d003 was 0.76 nm which meant that after modification tri-sodium citrate did not intercalate into the interlayer of LDHs. The result corresponded with the negative sign of zeta potential of citrate modification redispersion in RO water. These results were indicative of citrate ions that were present around LDHs surfaces after modification.

The appearance of raw LDHs and citrate-modified LDHs (bead milling 15 minutes) after 1 hour holding time was shown in Figure 5a. Tri-sodium citrate-modified LDHs showed improved stability, compared to non-treated LDHs. The citrate-modified LDHs in water was stable up to 7 days. As shown in Figure 5a, the average particle size of raw LDHs in water was around 40 μm, which was larger than citrate-modified by bead milling -176 nm and also corresponded with the FE-SEM results. The raw LDHs revealed the presence of large agglomerated particle with particle size more than 5 μm, whereas the citrate-modified LDHs by bead milling technique was found in small particles with particle around 100-500 nm.

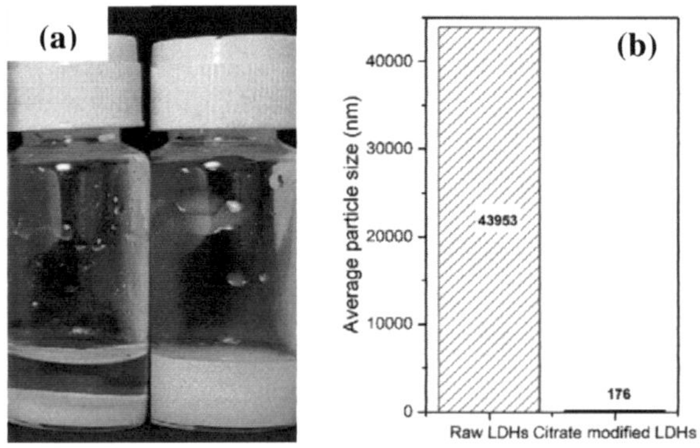

Figure 5. (a) appearance after 1 hour, (b) average particle size of Raw LDHs and citrate-modified LDHs by bead milling 15 minutes redispersion in water

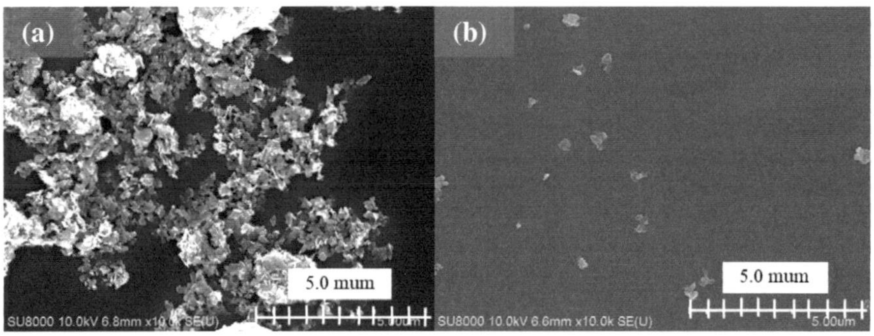

Figure 6. FE-SEM results of (a) Raw LDHs and (b) citrate-modified LDHs by bead milling 15 minutes (right) redispersion in water.

Overall, the bead milling technique reduced the agglomerated particle size and also induced the adsorption of citrate on LDHs particle. After modification of LDHs by tri-sodium, the stability of LDHs particle in water was improved.

4. Conclusion

In this study, the bead milling technique could disperse LDHs particles and induce adsorption of citrate on them. The optimum milling time was 15 minutes. However, the tri-sodium citrate attached on LDHs surface and did not intercalate into the interlayer of LDHs. The particle size of citrate-modified LDHs after redispersion in RO-water was 177 nm with stability >1 hour.

Acknowledgements

This work has partially been supported by Ministry of Science and Technology, Thailand, through its program of Center of Excellence Network and the SCG Chemicals Co., Ltd., for supporting the material used in this study. TC was supported by a Science Achievement Scholarship of Thailand.

5. References

[1] Evans D G and Duan X 2006 Preparation of layered double hydroxides and their applications as additives in polymers, as precursors to magnetic materials and in biology and medicine *Chem. Commun.* 485–96

[2] Velasco J I, Ardanuy M and Antunes M 2012 Layered double hydroxides (LDHs) as functional fillers in polymer nanocomposites *Woodhead Publishing Series in Composites Science and Engineering* ed F B T-A in P N Gao (Woodhead Publishing) pp 91–130

[3] Choy J H, Choi S J, Oh J M and Park T 2007 Clay minerals and layered double hydroxides for novel biological applications *Appl. Clay Sci.* **36** 122–32

[4] Pang X, Chen L, Liu Y, Chi M, Li Z and Plank J 2017 Growth behavior of water dispersed MgAl layered double hydroxide nanosheets † *RSC Adv.* **7** 14989–97

[5] Kim M S, Louis K M, Pedersen J A, Hamers R J, Peterson R E and Heideman W 2014 Using citrate-functionalized TiO2nanoparticles to study the effect of particle size on zebrafish embryo toxicity *Analyst* **139** 964–72

[6] Prathna T C, Chandrasekaran N and Mukherjee A 2011 Studies on aggregation behaviour of silver nanoparticles in aqueous matrices: Effect of surface functionalization and matrix composition *Colloids Surfaces A Physicochem. Eng. Asp.* **390** 216–24

[7] NETZSCH GmbH & Co. Holding KG. Wet grinding and dispersing with laboratory agitator bead mills of the Mini Series and Micro Series

[8] Montalescot V, Rinaldi T, Touchard R, Jubeau S, Frappart M, Jaouen P, Bourseau P and Marchal L 2015 Optimization of bead milling parameters for the cell disruption of microalgae: Process modeling and application to Porphyridium cruentum and Nannochloropsis oculata *Bioresour. Technol.* **196** 339–46

[9] Mandzy N, Grulke E and Druffel T 2005 Breakage of TiO2agglomerates in electrostatically stabilized aqueous dispersions *Powder Technol.* **160** 121–6

[10] Mahjoubi F Z, Khalidi A, Abdennouri M and Barka N 2015 Zn–Al layered double hydroxides intercalated with carbonate, nitrate, chloride and sulphate ions: Synthesis, characterisation and dye removal properties *J. Taibah Univ. Sci.* **11** 90–100

[11] Giri A, Makhal A, Ghosh B, Raychaudhuri A K and Pal S K 2010 Functionalization of manganite nanoparticles and their interaction with biologically relevant small ligands: Picosecond time-resolved FRET studies *Nanoscale* **2** 2704–9

IOP Conf. Series: Materials Science and Engineering **654** (2019) 012008 doi:10.1088/1757-899X/654/1/012008

Soot Oxidation Studies on $SrMn_{0.98}B_{0.02}O_3$ (B – Fe, Ni) Perovskites

Hari Prasad Uppara[1], Armaan Feroz[1], Nithin Samuel John[1], Sunit Kumar Singh[2], Nitin K Labhsetwar[2], Harshini Dasari[1]

[1] Department of Chemical Engineering, Manipal Institute of Technology, Manipal Academy of Higher Education, Manipal, 576104, India.
[2]Energy and Resource Management Division, CSIR-National Environmental Engineering Research Institute, Nagpur, Maharashtra, India.

E-mail address: harshini.dasari@manipal.edu

Abstract. In this paper, the performance of $SrMn_{0.98}B_{0.02}O_3$ (B = Fe, Ni) perovskites was evaluated for catalytic soot oxidation applications. All the samples exhibited a nano-crystalline hexagonal structure (XRD analysis) with the crystal sizes ranges 30 to 46 nm. XPS analysis reveals the presence of multiple valence states of Ni cations (Ni^{2+}/Ni^{3+}) and Mn cations (Mn^{4+}/Mn^{3+}) cations with the divalent strontium and trivalent iron cation on the surface of the perovskites. O_2-TPD analysis infers that the sample $SrMn_{0.98}Ni_{0.02}O_3$ showed the highest quantity of surface adsorbed oxygen molecules (78.87 μmol.g^{-1}) followed by $SrMnO_3$ sample (72.16 μmol.g^{-1}). Furthermore, the soot-TPR analysis confirms that the sample $SrMn_{0.98}Ni_{0.02}O_3$ shown the higher catalytic activity than $SrMn_{0.98}Fe_{0.02}O_3$due to synergetic effect between Mn - Ni cations, the higher reducibility tendency of the metal ions and the higher quantity of active oxygen species initiated the structural defects in the catalysts. The superior catalytic activity of soot oxidation of the as-synthesized samples was observed in the order of $SrMn_{0.98}Ni_{0.02}O_3$> $SrMn_{0.98}Fe_{0.02}O_3$>$SrMnO_3$>bare soot.

1. Introduction

Diesel engines hold higher fuel efficiency when compared to petrol engines but these engines contribute emissions of hydrocarbons, NOx and particulate matter. The particulate matter poses a serious threat to the environment and living beings [1]. Actually, the emitted soot from diesel engines oxidizes at above 600°C, but the operating engines (light to heavy-duty) are 150 – 350°C. Therefore, various catalytic processes such as diesel oxidation catalyst (DOC) [2] and diesel particulate filters (DPF) [3] have used to control the emissions. To operate within, the researchers need to explore novel catalysts to hold good thermal stability, chemically resistant and good catalytic properties. So far, many catalysts such as noble metal-based [4], ceria-based [5], and perovskite oxides [6] have studied for soot oxidation. However, the parameters like thermal stability, sensitive to be chemical poisoning and economic viability remain open [7]. Perovskites have used as a substitute for noble metal-based catalysts, which has the structure (ABO_3) and flexible in their structure. Perovskites have extensively explored because of their excellent oxidation activity. This is due to the creation of oxygen vacancies because of doping/substituting with other metals [8]. These oxygen vacancies allow more amount of gaseous oxygen to adsorb results in the enhancement of soot oxidation.

Doroftei et al [9] investigated properties of $SrMnO_3$ and applied for combustion of propane. The authors examined that 50% and 90% conversion of propane have observed at around 240°C and 350°C

Content from this work may be used under the terms of the Creative Commons Attribution 3.0 licence. Any further distribution of this work must maintain attribution to the author(s) and the title of the work, journal citation and DOI.
Published under licence by IOP Publishing Ltd

respectively. The superior catalytic behaviour of $SrMnO_3$ was due to redox properties of Mn (+4 and +3) and generation of a larger quantity of oxygen vacancy density (O^- and O^{2-}). Zhang et al [10] reported complete oxidation of CO over microspheres of binary oxides of CeO_2-MnO_2 composite at around 206°C. The superior catalytic performance of Ce-Mn oxide was due to higher surface area, more oxygen storage capacity and easy redox properties. Rezlescu et al [11] reported $SrMn_{1-x}Ce_xO_3$ (x= 0-0.2) perovskite, studied the surface properties of the samples and tested for combustion of dilute acetone. They examined that $SrMn_{0.9}Ce_{0.2}O_3$ took advantage over $SrMnO_3$ because of the formation of larger active species (O^-, O^{2-}), generated due to redox properties of Ce ions. In the current work, the $SrMn_{0.98}B_{0.02}O_3$ (B= Fe, Ni) perovskites were synthesized and applied for soot oxidation applications. The physio-chemical properties were examined using XRD, BET surface area, SEM, XPS, O_2-TPD, and Soot-TPR. The soot oxidation experiments on $SrMn_{0.98}B_{0.02}O_3$ (B = Fe, Ni) were carried out in a TGA instrument. The detailed soot oxidation studies on these samples have not studied and this prompted to undertake the soot oxidation investigations.

2. Experimentation

2.1 Preparation of perovskites

The three perovskites $SrMnO_3$ (SM), $SrMn_{0.98}Fe_{0.02}O_{3\pm\delta}$ (SMFe2%),, $SrMn_{0.98}Ni_{0.02}O_{3\pm\delta}$ (SMNi2%) were prepared using the ultra-sonicated sol-gel method. All the materials are of analytical grade. The dissolved metal nitrates in distilled water were ultra-sonicated for 60 min, followed by addition of ammonia solution to metal nitrates mixture to maintain the pH ~10 and kept for constant stirring at 500 rpm and 90°C. The combustible agent citric acid was added to the resultant mixture, maintaining the molar ratio of nitrates and citric acid as 1:1.5. The final mixture has kept for constant stirring at 90°C until it forms the gel. The formed gel transferred into the oven at 120°C for overnight. Thus, the combusted sample kept for calcination in a furnace at 950ºC for 24 h.

2.2. Characterization and performance of catalytic studies

The structural analysis of the as-synthesized samples were done using XRD technique (Rigaku) with Cu Kα radiation (λ =1.54 Å, 40 kV, 20 mA). The BET surface area of the samples was examined using SmartSorb 92/93 BET surface area analyzer. SEM analysis was performed on Carl Zeiss (Germany) to understand the surface morphological properties of the samples. O_2-TPD experiments were performed on Micromeritics Auto Chem II 2920 instrument to examine the active oxygen species available on the surface of catalysts. XPS analysis was carried on XPS microprobe (ESCALAB Xi^+) equipped with (Mono 900 μm and Al-Kα (1486.6 eV)). Soot-TPR experiments were carried out in the TGA instrument in the temperature range of 100 to 800° C in the absence of bulk oxygen species, performed in Ar medium (99.99% UHP). In soot oxidation experiments, typically 15 mg of the tightly contacted catalyst/ Printex carbon black mixture (1:10) in TGA instrument. All the catalysts have tested for soot oxidation reactions in the temperature range of 50 to 700°C at the ramp input of 10°C per min. The CO_2 gas at the outlet of TA instrument have recorded using ACCURA CO_2 analyzer.

3. Results and Discussion

3.1 Structural and surface morphological studies:

The structural analysis of as-prepared perovskites are studied using XRD technique, and the spectra are depicted in figure 1 (i) which exhibits hexagonal phase and the characteristic peaks are well accord with standard data ICDD: 00-024-1213 (space group - P63/mmc). The introduction of the foreign metal cation (B^{n+}) into the crystal lattice of SM sample, contributes a slight shift in (110) peak towards higher angles, which confirms the contraction of the crystal lattice of the samples. In addition, the lattice parameters of these doped samples (shown in Table 1) is smaller than the $SrMnO_3$, which might be associated with the partial replacement of foreign metal cation (B^{n+}) at the B-site of $SrMnO_3$ (the ionic radii are Sr^{2+} (1.18 Å), Mn^{3+} (0.65 Å), Mn^{4+} (0.53 Å), Fe^{3+} (0.645 Å), Ni^{2+} (0.69 Å), Ni^{3+} (0.56 Å) respectively) [12]. An extra weak signal was observed for the sample SMNi2% at around 32.16°, indicated slightly perturb the crystallinity of the sample or reached the maximum solubility of the dopant. The BET surface area of all the catalysts is observed in the range of 1.58 to 3.06 m^2/g. SEM images of the perovskites are shown in figure 1 (ii). All the prepared catalysts exhibited nano-crystals with non-uniform distribution of particles sizes.

Table 1: Structural parameters estimated from XRD analysis.

Perovskite	Crystal Size, L (nm)	Lattice Parameters a (Å)	b (Å)	c (Å)	Volume ($Å^3$)
SM	46.6	5.468	5.468	9.105	235.82
SMFe2%	29.9	5.467	5.467	9.099	234.90
SMNi2%	35.8	5.446	5.446	9.066	232.96

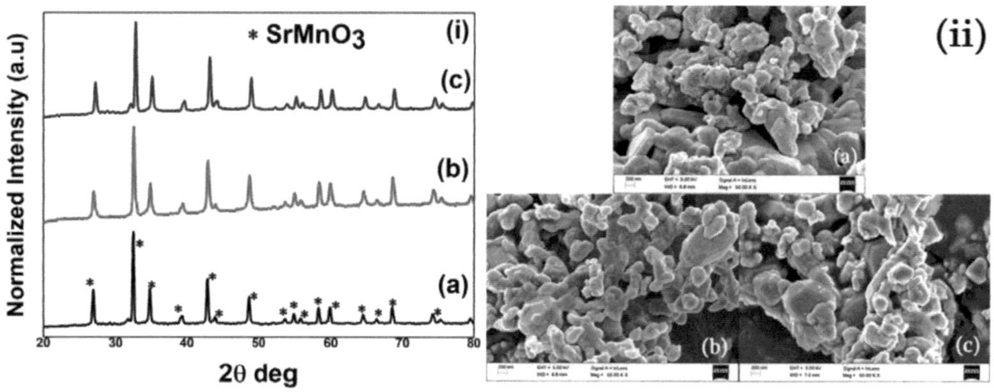

Figure 1. (i) XRD pattern (ii) SEM images (a) SM, (b) SMFe2% and (c) SMNi2%.

3.2 XPS analysis

Fe2p spectra is depicted in figure 2 (1-6). The peaks of Fe2p of the sample SMFe2%, cornered at around 711.9 eV and 725.8 eV are ascribed to Fe^{3+} 2p3/2 and Fe^{3+} 2p1/2 with the spin-orbital energy difference of 13.9 eV [13]. A very small super-imposed shoulder peak of Fe^{4+} cation is also observed at around ~712 eV. In addition, the shake-up satellite peaks of Fe^{3+} spin orbitals are also observed at around 719.1 eV and 733.3 eV. The Ni2p peaks (SMNi2%) are shown in figure 2 (7-12). The two spin-orbital doublets are observed on the surface of the catalyst. The first doublet binding energies (850.0 eV, 869.5 eV) and the second doublet energies (855.7 eV, 873.3 eV) are related to Ni^{2+} and Ni^{3+} cations respectively [14]. Furthermore, the two shake-up satellite peaks are also observed at 860.97 eV is related to Ni^{2+} and the peak cornered at 865.6 eV is ascribed to Ni^{3+} cation respectively, these peaks are consistent with the reported literature [15]. In this study, the sample SMNi2% showed more predominant Ni^{3+} cations on the surface. Figure 2(13-16) depicts Mn2p spectra of the samples. For the un-doped SM sample, there are two main peaks observed at ~641.3 eV and ~643.2 eV with the corresponding spin-orbital split at around ~652.8 eV and ~654.6 eV is related to Mn^{3+} and Mn^{4+} respectively, and these B.E's are well-matched with the literature [9, 16, 17]. Furthermore, the Mn2p peaks for the doped samples are analyzed and reported Mn^{3+} at (SMFe2% - 642.1 eV, 653.6 eV; SMNi2% - 642.3 eV, 653.8 eV) and Mn^{4+} at (SMFe2% - 643.9 eV, 655.5 eV; SMNi2% - 644.2 eV, 655.7 eV) respectively. For the doped sample, the peaks are shifted towards higher B.E's when compared with the un-doped sample because of partial substitution of more electronegative elements (Fe-1.83, Ni-1.91) with the less electronegative element (Mn-1.55). Figure 2(17-19) shows the O1s peaks of the as-prepared samples. It can be noted that the release of O_2^-/O^- is more crucial to carry out the soot oxidation reactions at lesser temperatures [18]. As summarized in Table 2, the binding energies of the surface adsorbed oxygen species (O_2^- /O-) of the samples SM, SMFe2% and SMNi2% are cornered at around 530.0, 530.7 and 531.0 eV respectively and the corresponding calculated percentages of surface adsorbed oxygen species are 48.1%, 42.6%, and 48.2% respectively. Thus, the sample SM and SMNi2% exhibited the almost the same quantity of active oxygen species, indicating both the sample might exhibit better soot oxidation rate when compared with the SMFe2% sample.

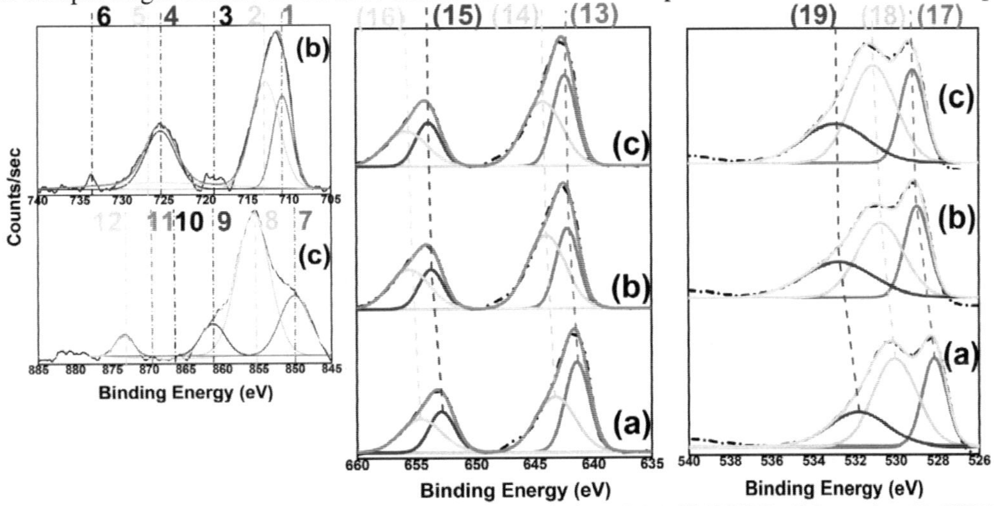

Figure 2. XPS peaks of the samples (a) SM, (b) SMFe2% and (c) SMNi2%. [Note: 1 – Fe(III)$2p_{3/2}$, 2 – Fe(IV)$2p_{3/2}$, 3 – Fe(III)sat, 4 – Fe(III)$2p_{1/2}$, 5 – Fe(IV)$2p_{1/2}$, 6 – Fe(IV)sat ,7 – Ni (II) $2p_{3/2}$, 8 – Ni (III) $2p_{3/2}$, 9 – Ni (II)sat, 10 – Ni (III)sat, 11 – Ni (II) $2p_{1/2}$, 12 – Ni (III) $2p_{1/2}$, 13 – Mn(III) $2p_{3/2}$, 14 – Mn(IV) $2p_{3/2}$, 15 – Mn(III) $2p_{1/2}$, 16 – Mn(IV) $2p_{1/2}$, 17- O^{2-}, 18 - O^-/O_2^-,, 19 - O_{hyd}]

3.3 O_2-TPD and soot-TPR analysis:

O_2-TPD peaks are divided into three regions and the corresponding graphs are shown in figure 3 (i). The three peaks at below 200°C, 200-600°C and at above 600°Care ascribed to weakly adsorbed oxygen, surface adsorbed (α_S - O^-,O_2^-) and lattice oxygen species (β_L - O^{2-}) respectively. For the SM sample, the two detected desorption peaks (331°C and 572°C) are ascribed to α_S - species with the calculated amount of desorbed oxygen is 72.16 μmol.g^{-1}. Similarly, the detected α_S desorbed oxygen species for the SMNi2% sample is at around 593°C, with the corresponding amount of oxygen species is 78.87 μmol.g^{-1}. However, the α_S - desorbed oxygen species is not detected for SMFe2% sample. Moreover, the calculated amount of β_L desorbed oxygen molecules for the samples SM, SMFe2% and SMNi2% are 6.57, 53.63 and 57.64 μmol.g^{-1} respectively. The sample SMNi2% showed the highest quantity of 'α_S' when compared with the other samples, the results are consistent with the O1s analysis (figure 2 (17-19)). Figure 3 (ii) depicts the soot-TPR analysis, to further investigate the presence of active species and the reducibility of the catalysts in the presence of soot. In the Soot-TPR experiments, the soot can only be oxidized because of reduction of catalysts which successively enhances the release of the surface adsorbed oxygen species (O_2^- /O^-) [19]. The catalysts reduction by soot at low temperatures observed the sequence as follows SMNi2%>SMFe2%>SM, the trend of the catalytic property of these catalysts is well consistent with the XPS analysis.

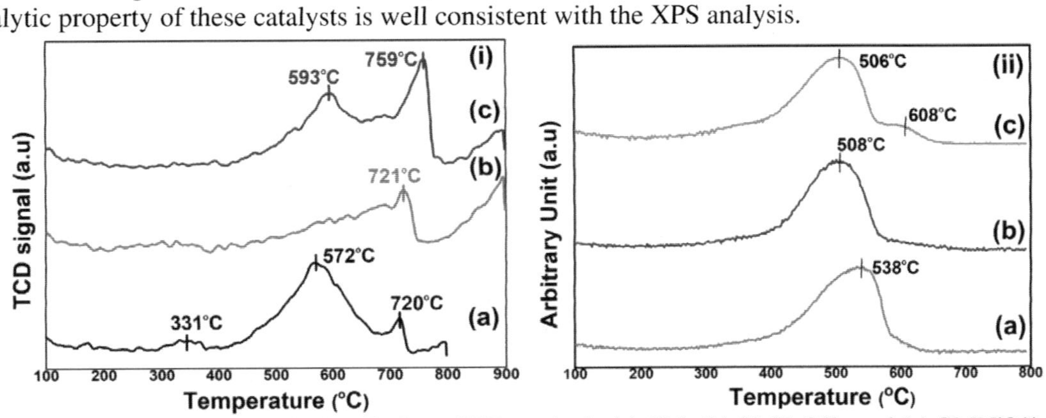

Figure 3. (i) O_2-TPD analysis (ii) Soot-TPR analysis (a) SM, (b) SMFe2%, and (c) SMNi2%.

3.4 Catalytic activity studies:

Figure 4 (i) depicts the conversion profiles of un-catalyzed and catalyzed soot in a temperature range of 150 to 700°C, the corresponding soot conversion temperatures are displayed in table 2. The $T_{50\%}$ values of the samples un-catalyzed soot, SM, SMFe2% and SMNi2% are found to be 614, 458, 445, 438°C respectively. The $T_{50\%}$ conversion of the un-catalyzed soot is at the temperature of around 614°C, whereas for the pure perovskite sample $SrMnO_3$ takes place at around 458°C which means calculated decrement of temperature is around 156°C. Further decrement of soot conversion temperature is observed for doping of Ni and Fe (2 mole %) at B-site of $SrMnO_3$. The $T_{50\%}$ of the corresponding SMFe2% and SMNi2% catalysts is lowered by ca. 169°C and 176°C respectively when compared with the un-catalyzed soot (614°C). It is very interesting to note that SMNi2% exhibited highest soot oxidation rate compared with the SMFe2% and SM sample. And the results are consistent with the soot-TPR analysis as noticed earlier, due to the creation of more oxygen defects in the perovskite. The soot conversion results are further confirmed by recording CO_2 concentration at the exit of the TGA instrument, the CO_2 concentration (%) v/s temperature as depicted in figure 4 (ii). The maximum soot conversion was observed at 628°C, 477°C, 458°C, 447°C for the samples bare soot, SM, SMFe2% and SMNi2% respectively. Thus, the CO_2 conversion profile is also consistent with the results obtained from soot-TPR and the soot conversion profiles.

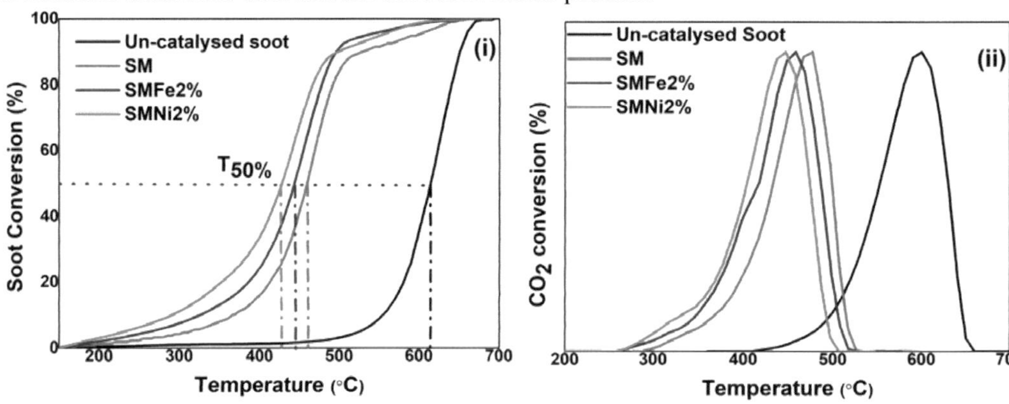

Figure 4. (i) Soot oxidation activity (ii) the CO_2 conversions (%) (a) SM, (b) SMFe2% and (c) SMNi2%.

Table 2. XPS – B.E's of O1s, surface metal ions and surface oxygen species from O_2-TPD and the corresponding soot conversion temperatures.

Sample	Binding Energies (eV)			[a]Os (%)	Atomic (%) Mn^{4+}/(Total)	[b]α_S (μmol. g^{-1})	[c]β_L (μmol. g^{-1})	$T_{50\%}$	T_{max}
	O_L	O_S	O_{hyd}						
SM	528.1	530.0	531.8	48.1	52.7	72.16	6.57	456	477
SMFe2%	528.9	530.7	532.7	42.6	61.5	[d]nd	53.63	445	458
SMNi2%	529.1	531.0	532.9	48.2	55.9	78.87	57.64	438	447

[a]Surface adsorbed oxygen (%) = $\frac{O_S}{O_L+O_S+O_{hyd}}$.

[b]Amount of surface adsorbed oxygen species from O_2-TPD analysis in between 200-600°C.

[c]Amount of lattice oxygen molecules desorption from O_2-TPD at above 600°C; [d]no detection.

4. Conclusions

$SrMn_{0.98}B_{0.02}O_3$ (B = Fe, Ni) perovskite-type catalysts have successfully synthesized and applied for the soot oxidation applications. The XRD patterns showed hexagonal structure. XPS results evidenced the Sr^{2+}, Fe^{3+} and redox coupling properties of Mn^{4+}/Mn^{3+} and Ni^{3+}/Ni^{2+} respective cations on the surface. Besides, the calculated quantity of release of surface adsorbed oxygen species from the surface of SM, SMFe2% and SMNi2% are 48.1, 42.6 and 48.2% respectively. Furthermore, the O_2-TPD experiments evidenced that the mobility of surface adsorbed oxygen molecules is highest for SMNi2% (78.87 µmol.g-1). From soot-TPR experiments, SMNi2% exhibited a lesser soot-catalyst reduction temperature compared with the other perovskites. Moreover, the sample SMNi2% showed more oxygen vacancy density than the SMFe2%. Thus, the sample SMNi2% exhibited superior catalytic activity ($T_{50\%}$=438°C).

Acknowledgments

This work was also supported by the KIST Institutional Program (Project No. 2Z05800)."

5. References

[1] J. Adler, Int. J. Appl. Ceram. Technol. 2 (2005) 429–439.
[2] J.A. Onrubia, B. Pereda-Ayo, U. De-La-Torre, J.R. González-Velasco, Appl. Catal. B Environ. 213 (2017) 198–210.
[3] D. Fino, S. Bensaid, M. Piumetti, N. Russo, Appl. Catal. A Gen. 509 (2016) 75–96.
[4] J. Oi-Uchisawa, A. Obuchi, R. Enomoto, S. Liu, T. Nanba, S. Kushiyama, Appl. Catal. B Environ. 26 (2000) 17–24.
[5] K. Krishna, A. Bueno-López, M. Makkee, J.A. Moulijn, Appl. Catal. B Environ. 75 (2007) 189–200.
[6] Y. Da, L. Zeng, C. Wang, T. Mao, R. Chen, C. Gong, G. Fan, Catal. Today 327 (2019) 73–80.
[7] L. Urán, J. Gallego, W.Y. Li, A. Santamaría, Appl. Catal. A Gen. (2019) 157–169.
[8] S. Royer, D. Duprez, F. Can, X. Courtois, C. Batiot-Dupeyrat, S. Laassiri, H. Alamdari, Chem. Rev. 114 (2014) 10292–10368.
[9] C. Doroftei, L. Leontie, RSC Adv. 7 (2017) 27863–27871.
[10] J. Zhang, Y. Cao, C.A. Wang, R. Ran, ACS Appl. Mater. Interfaces 8 (2016) 8670–8677.
[11] N. Rezlescu, E. Rezlescu, P.D. Popa, C. Doroftei, M. Ignat, Mater. Chem. Phys. 182 (2016) 332–337.
[12] R.D. Shannon, C.T. Prewitt, Acta Crystallogr. Sect. B Struct. Crystallogr. Cryst. Chem. 25 (1969) 925–946.
[13] L. Chen, G. Liu, N. Feng, J. Yu, J. Meng, F. Fang, P. Zhao, L. Wang, Appl. Surf. Sci. 467–468 (2019) 1088–1103.
[14] L. Mi, W. Wei, S. Huang, S. Cui, W. Zhang, H. Hou, W. Chen, J. Mater. Chem. A 3 (2015) 20973–20982.
[15] Z. Ren, Z. Wu, W. Song, W. Xiao, Y. Guo, J. Ding, S.L. Suib, P.X. Gao, Appl. Catal. B Environ. 180 (2016) 150–160.
[16] S. Li, S. Yan, Y. Xia, B. Cui, Y. Pu, Y. Ye, D. Wang, Y.Q. Liu, B. Chen, Appl. Catal. A Gen. 570 (2019) 299–307.
[17] X. Lin, S.S. Li, H.H. He, Z. Wu, J. Wu, L. Chen, D. Ye, M. Fu, Appl. Catal. A Gen. 223 (2018) 181–189.
[18] D. Mukherjee, B.G. Rao, B.M. Reddy, Appl. Catal. B Environ. 197 (2016) 105–115.
[19] H.P. Uppara, H. Dasari, S.K. Singh, N.K. Labhsetwar, M.S. Murari, Catal. Letters (2019).

IOP Conf. Series: Materials Science and Engineering **654** (2019) 012009 doi:10.1088/1757-899X/654/1/012009

The Preparation of Highly Transparent Layered Double Hydroxides (LDHs)-Poly(Methyl Methacrylate) (PMMA) Nanocomposites Sheets by Batch Cell Polymerization Method

K Joonlapak[1], P Wiwattanakull[1], T Jirathampradhab[1], T Akhadejdamrong[4], A Wongariyakawee[4] and T Srikhirin[1, 2, 3]

[1] School of Materials Science and Innovation, Faculty of Science, Mahidol University, 999 Phutthamonthon Sai 4 Road, Salaya, Nakhon Pathom 73170, Thailand
[2] Department of Physics, Faculty of Science, Mahidol University, Bangkok, 10400, Thailand
[3] NANOTEC Center of Excellence, Mahidol University, Bangkok, 10400, Thailand
[4] SCG chemicals, Siam Cement Road, Bang Sue, Bangkok, 10800, Thailand

toemsak.sri@mahidol.ac.th

Abstract. Enhancement polymer properties using various clay nanocomposites has been interesting in polymer technology. Herein, a preparation of transparent layered double hydroxides/poly methylmethacrylate (LDHs/PMMA) nanocomposites sheet was achieved by batch cell polymerization. Nanometre-sized MgAl-LDHs in toluene showed well dispersed suspension and was translucent, indicating delamination of LDHs stacks. To prepare PMMA-LDHs nanocomposites, the delaminated MgAl-LDHs in toluene was mixed with methymethacrylate oligomer in order to form slurry of dispersed LDHs in PMMA pre-polymer. The LDHs/PMMA nanocomposites sheet was subsequently prepared using batch cell polymerization at 60oC for 5 hours in water bath and it was then post curing at 80oC for 30 minutes. The optical properties of MgAl-LDHs/PMMA nanocomposites were strongly depending upon a dispersion behaviour of MgAl-LDHs.

1. Introduction

Transparent plastic is material that exhibits transmission of visible light greater than 80% such as polycarbonate (PC), polystyrene (PS) polyethylene terephthalate (PET) and polymethyl methacrylate (PMMA). Transparent plastics are increasingly used to replace glass as construction material because of the ease of fabrication, light weight, safety and cost effective. In fact, the mechanical properties of PMMA, such as stiffness and strength, are much lower than those of metals. However, its mechanical properties are generally improved by applying reinforcement materials such as clay minerals (i.e. montmorillonite, kaolinite, layered double hydroxides). The clay mineral, especially clay nanoparticles, have been studied as a filler in polymeric matrix to enhance mechanical property while maintaining the transparency of the polymer.

The layered double hydroxides (LDHs) is a synthetic clay material containing inorganic layers (layer domain) and the interlayer anions, having the following chemical formula; $[M^{z+}_{1-x}M^{y+}_{x}(OH)_2]^{a+}(X^{n-})_{a/n}{\cdot}mH_2O$ (M^{z+}, z = 1or 2; M^{y+}, y = 3 or 4). Due to its structure, the LDHs are adaptively used for several applications such as catalysts [1,2] , waste absorber [3] , UV/ IR absorbers [4–6] , flame retardants[7–10] , drugs release [11], and nano-fillers in polymer. A study of LDHs as

Content from this work may be used under the terms of the Creative Commons Attribution 3.0 licence. Any further distribution of this work must maintain attribution to the author(s) and the title of the work, journal citation and DOI.
Published under licence by IOP Publishing Ltd

filler for polymer is widely reported in literatures [10]. The polymer/LDHs nanocomposites show an improvement of mechanical and thermal properties over pristine polymer. It is known that nanometer size additives do not affect optical property of polymer. LDHs normally are hydrophilic and cannot be dispersed in nonpolar solvents and are not compatible with non-polar polymer. Hydrophobicity of LDHs can be altered by intercalating with long chain anionic surfactants (e.g. dodecyl sulfate (DDS). Anionic surfactant can reduce IR absorption and/or flame retardant properties of LDHs. Thus, the LDHs intercalated with anions such as carbonate and borate was studied as additive in polymer [12].

One of simple method mostly used to prepare LDHs/polymer nanocomposites is melt mixing technique. LDHs powder and polymer is mechanically mixed at high temperatures (e.g. 160–200°C). However, dried LDHs powder generally forms aggregate leading poor dispersion in the polymer matrix. This phenomenon can affect the properties of the polymer. Therefore, in this work, LDHs/PMMA nanocomposites were fabricated by mixing MMA oligomer and carbonate intercalated MgAl-LDHs via simple solvent mixing method.

2. Experimental

2.1 Materials

Methyl methacrylate (MMA) was used in an oligomer form. The Mg/Al layered double hydroxides (MgAl-LDHs) powder was supported by SCG Chemicals Co., Ltd., Thailand. 2,2'-Azobisisobutylonitrile(AIBN), commercial grade was purchased from Otsuka chemical Co.,Ltd. Acetone(CH_2COCH_3), Methanol(CH_3OH), Ethanol(C_2H_6O), Iso-propanol(C_3H_8O), Hexane(C_6H_{14}) and toluene(C_7H_8) analytical grade was purchased from RCI labscan (Thailand).

Glass molds should be glossy and should be inspected for defects such as scratch on the surface before use. The PVC gasket stay between two glass molds that look like a sandwich for control thickness of PMMA sheet by using C-clamps.

2.2 Dispersion of Mg/Al LDHs

10%wt/v of MgAl-LDHs powder was prepared by dispersing the Mg/Al LDHs powder in organic solvent with different polarity for their dispersion. The mixture was continuously stirred until it was seen as stable solution. Care must be in order to keep the amount of solvent to be as low as possible.

2.3 Preparation of LDHs/PMMA nanocomposites sheets

The experimental conditions were as follow. The first stage, composites solution, the (200 ml) MMA-oligomer (MMA syrup) was slowly added into 20 ml of MgAl-LDHs (from step dispersion of Mg/Al LDHs) and the solution was vigorously stirred for 30 min. After that 0.4 g AIBN as initiator was added into mixture and then continuously stirred for 30 min. Vacuum process was acquired in order to eliminate bubbles, produced in composites solution. In the second stage, casting polymerization, the composites solution was injected into glass mold having 4 mm spacer. The mold was put into water bath at 60°C for 5 hrs and post curing at 80°C for 30 min. Finally, the glass mold was slowly cooled down to room temperature and LDHs/PMMA nanocomposites sheet was taken out from mold to obtain 1 % wt./v of the LDH (1%LDHs).

2.4 Characterizations

The optical transparency of nanocomposite sheet was observed by UV-Vis spectrophotometer (ISR-2600Plud integrating Sphere Attachment, UV-2600, Shimadza) with transmission mode in range of 200-800 nm.

3. Result and discussion

The behaviour of MgAl-LDHs dispersions are summarised in Table 1. The criterions that used to pick intermediate solvent for preparing LDHs/PMMA nanocomposites sheet are miscibility with MMA oligomer and good dispersibility of MgAl-LDHs. For example, acetone and toluene are good solvent for MMA oligomer and good dispersant for LDHs indicated by thick layer of LDHs. However, both exhibit different LDHs appearance. The LDHs layer was translucent and pearlescent in toluene, whereas there was opaque in acetone as shown in Figure1.

Table 1 Observations on selected MgAl-LDHs dispersions

Polarity index	Dispersion solvent	Apparent of LDHs	Miscibility with MMA oligomer
9	Water	Opaque	✗
6.6	Methanol	Opaque	✗
5.4	Acetone	Opaque	✓
5.2	Ethanol	Opaque	✗
4.3	Iso-propanol	Opaque	✗
2.3	Toluene	Translucent	✓
0	Hexane	Opaque	✗

MgAl-LDHs was dispersed in organic solvents with different polarity. All of them showed either precipitated product or no dispersion. It was found that the LDHs in toluene showed stable dispersion as translucent and pearlescent suspension. This was due to the dispersion of LDHs in toluene exhibited a match in the surface of MgAl-LDHs and the solvent polarity, leading of Tyndall light scattering effect. However, the LDHs in acetone and DI water showed poor dispersion as white solid suspension indicating neither types of dispersion is stable for LDHs loading. Hence, the LDHs dispersion in acetone and DI water could not be transparent phase and be seen like aggregates form. The behaviour for these MgAl-LDHs dispersions affected optical properties of nanocomposites sheets as showed in figure 2. It should be pointed out that the better stability and degree of transparency in MMA oligomers, the better the transparency that could be obtained.

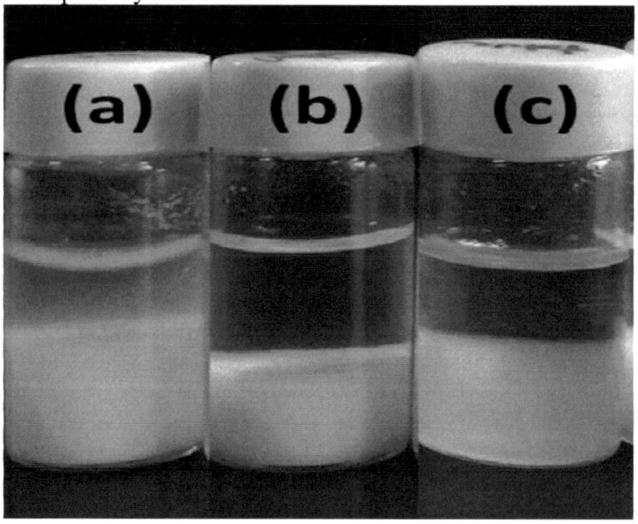

Figure 1 illustrates photographs of LDHs nanoparticles in (a) DI water, (b) Acetone and (c) toluene

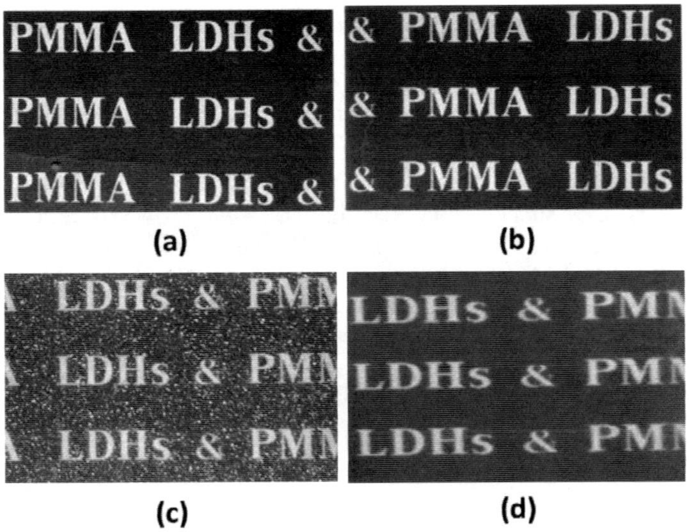

Figure 2. Appearance of LDHs/PMMA nanocomposites sheets a) neat PMMA, b) 1%LDHs in toluene, c) 1%LDHs in acetone and d) 1%LDHs without solvent

Figure 2 revealed that LDHs/PMMA nanocomposites with toluene as solvent showed no significant difference with neat PMMA when observed with naked eye. They were transparent and clear. However, opaque LDHs/PMMA nanocomposites sheet were observed in case of powder sample (without solvent) and with acetone as solvent. This was due to poor dispersion of LDHs in PMMA matrix. UV-Vis transmission spectra of LDHs/PMMA nanocomposites sheet according to samples in Figure 2 under UV-Visible wavelength were showed as seen in Figure 3.

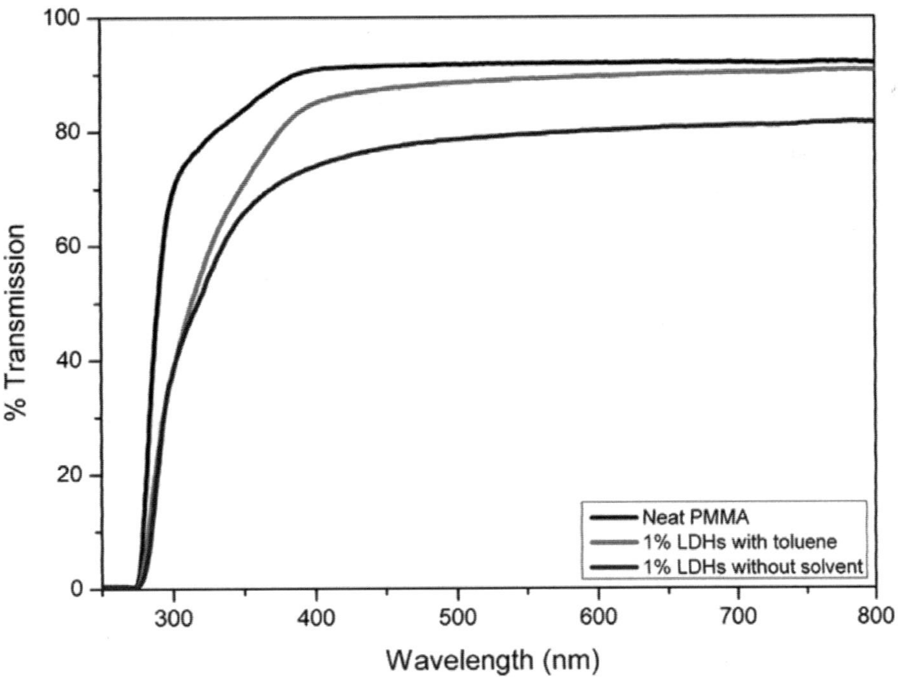

Figure 3. UV-Vis transmission spectra of the LDHs/PMMA nanocomposites with 1% LDHs contents with and without toluene.

Considering optical property of LDHs/PMMA nanocomposites sheet, transmission of each sample at visible light region was determined. Neat PMMA (without LDHs) offered the highest transmission. It could give 91%T. Although transmission of visible light was slightly decreased in sample prepared with toluene, it was still high as 89%T at 1% LDHs loading. Highly optical clarity of LDHs/PMMA nanocomposites sheet was due to disorderly exfoliated morphology leading the LDHs nanoparticles in PMMA matrix were relatively smaller than wavelength of visible light [13]. However, the LDHs/PMMA nanocomposites sheet prepared without solvent showed dramatic decrease of %T as compared to those with toluene. %T of the LDHs/PMMA nanocomposites sheet without toluene were approximately 79% at 1% LDHs loading. This was due to the LDHs nanoparticles were not exfoliated and tended to highly aggregated in PMMA matrix as shown in figure 2.

4. Conclusion

The LDHs/PMMA nanocomposites with MgAl-LDHs were successfully fabricated by solvent mixing process. LDHs nanocomposites dispersed in toluene slightly affected transparency of PMMA. However, LDHs contents caused the agglomeration of LDHs layers resulting in adverse effects. Consequently, we believe this process has provided great potential as simple and convenient method for fabrication of LDHs/PMMA nanocomposites sheet.

Acknowledgements

This work has partially been supported by SCG Chemicals Co., Ltd., Thailand for supporting the material used in this study. The authors are also thankful to the Materials Science and Engineering Program at Mahidol University, Thailand and Thailand Center of Excellence in Physics at Chiang Mai University, Thailand

5. References

[1] Wei M, Zhao Y, Wang Z L, Duan X and Lu J 2009 Biotemplated Hierarchical Nanostructure of Layered Double Hydroxides with Improved Photocatalysis Performance *ACS Nano* **3** 4009–16

[2] Zhao Y, Zhang S, Li B, Yan H, He S, Tian L, Shi W, Ma J, Wei M, Evans D G and Duan X 2011 A family of visible-light responsive photocatalysts obtained by dispersing CrO 6 octahedra into a hydrotalcite matrix *Chem. - A Eur. J.* **17** 13175–81

[3] Prasanna S V., Rao R A P and Kamath P V 2006 Layered double hydroxides as potential chromate scavengers *J. Colloid Interface Sci.* **304** 292–9

[4] Evans D G and Duan X 2006 Preparation of layered double hydroxides and their applications as additives in polymers, as precursors to magnetic materials and in biology and medicine *Chem. Commun.* 485–96

[5] Zhao X, Zhang F, Xu S, Evans D G and Duan X 2010 From layered double hydroxides to ZnO-based mixed metal oxides by thermal decomposition: Transformation mechanism and UV-blocking properties of the product *Chem. Mater.* **22** 3933–42

[6] Feng Y, Li D, Wang Y, Evans D G and Duan X 2006 Synthesis and characterization of a UV absorbent-intercalated Zn-Al layered double hydroxide *Polym. Degrad. Stab.* **91** 789–94

[7] Nyambo C, Songtipya P, Manias E, Jimenez-Gasco M M and Wilkie C A 2008 Effect of MgAl-layered double hydroxide exchanged with linear alkyl carboxylates on fire-retardancy of PMMA and PS *J. Mater. Chem.* **18** 4827–38

[8] Nyambo C, Wang D and Wilkie C A 2009 Will layered double hydroxides give nanocomposites with polar or non-polar polymers? *Polym. Adv. Technol.* **20** 332–40

[9] Manzi-Nshuti C, Wang D, Hossenlopp J M and Wilkie C A 2008 Aluminum-containing layered double hydroxides: The thermal, mechanical, and fire properties of (nano)composites of poly(methyl methacrylate) *J. Mater. Chem.* **18** 3091–102

[10]Gao Y, Wu J, Wang Q, Wilkie C A and O'Hare D 2014 Flame retardant polymer/layered double hydroxide nanocomposites *J. Mater. Chem. A* **2** 10996–1016

[11]Del Hoyo C 2007 Layered double hydroxides and human health: An overview *Appl. Clay Sci.* **36** 103–21

[12]Wang Q, Zhang X, Zhu J, Guo Z and O'Hare D 2012 Preparation of stable dispersions of layered double hydroxides (LDHs) in nonpolar hydrocarbons: New routes to polyolefin/LDH nanocomposites *Chem. Commun.* **48** 7450–2
[13]Wang G A, Wang C C and Chen C Y 2006 The disorderly exfoliated LDHs/PMMA nanocomposites synthesized by in situ bulk polymerization: The effects of LDH-U on thermal and mechanical properties *Polym. Degrad. Stab.* **91** 2443–50

IOP Conf. Series: Materials Science and Engineering **654** (2019) 012010 doi:10.1088/1757-899X/654/1/012010

The Structure and Properties of Sn/SiSn-nanodisperse Alloy Thin Films

Xiangdong Jiang[1], Mingcheng Li[1], Yangzhou Ming[1], Jimin Wang[1], and Wei Li[1]

[1]University of Electronic Science and Technology of China, No.4, Section 2, North Jianshe Road, Chengdu, 610054, China

E-mail: xdjiang@uestc.edu.cn

Abstract. Near-infrared brain imaging technology has great advantages in brain imaging and inspection of brain disorders compared to traditional brain imaging technology. Silicon-tin (SiSn) alloys are expected to be the material for infrared brain imaging detectors. The structure and properties of the SiSn alloy thin films with relatively low Sn concentration, which are the key for it to be used in near-infrared brain imaging technology, have not been reported yet. Here, we report the deposition time, growth temperature, microstructure, resistivity, and transmittance of amorphous silicon-tin (a-$Si_{1-x}Sn_x$) alloy thin films prepared by radio frequency (RF) magnetron sputtering. Reasonable deposition time and growth temperature for the preparation of the films are given in this paper. Sn nanocrystals are observed in the a-$Si_{1-x}Sn_x$ alloy thin films. The variations in resistivity and transmittance indicate that it has excellent electrical and optical properties so that it can be used as a near-infrared brain imaging detector material.

1. Introduction

Si remains a mainstream semiconductor material for numerous nanoelectronics, sensor and photovoltaic devices [1]. The alloying of Si with other IV group elements (SiGe [2], SiSn [3]-[5], SiGeSn [6]) is one way for apply Si to silicon-based devices because of their promising possibilities for band gap engineering and structural modification. Due to the diffusion process of tin (Sn) into the silicon lattice is low cost, scalable and manufacturable [7], the SiSn binary alloys are attracting considerable attention. At present, the preparation methods of SiSn alloy thin films mainly include co-evaporation technology [8], liquid phase epitaxy technology [9], ion beam induced epitaxial crystallization technology [10], chemical vapor deposition technology [11], magnetron sputtering [12] and molecular beam epitaxy [13]. The authors of [14] report the growth of epitaxial SiSn films with a Sn concentration from 10 to 35% and study the morphology and surface structure of the SiSn layers. The authors of [15] report the optical properties of such co-evaporated, nonhydrogenated amorphous silicon-tin (a-$Si_{1-x}Sn_x$) alloys, with x varying from 0 to 0.5. In the paper describing of application of radio frequency (RF) sputtering to obtain SiSn films, it report the structural, electrical, and optical properties of rf-sputtered a-$Si_{1-x}Sn_x$:H films in the composition range $0 \leq x \leq 0.51$ [16]. The optical band gap of SiSn varies linearly with increasing Sn concentration, and many research efforts have focused on applying SiSn alloys to the mid-far infrared band by increasing the concentration of doped Sn. The near-infrared (NIR) region of 700-1700nm, due to its long tissue penetration depths and ultralow tissue backgrounds [17], [18], has great application value in NIR brain imaging. In this paper, we report the a-$Si_{1-x}Sn_x$ alloy thin films prepared by magnetron sputtering at relatively low concentration of Sn, and show that Sn is dispersed in a-$Si_{1-x}Sn_x$ alloy thin films in the form of nanocrystals. Further,

Content from this work may be used under the terms of the Creative Commons Attribution 3.0 licence. Any further distribution of this work must maintain attribution to the author(s) and the title of the work, journal citation and DOI.
Published under licence by IOP Publishing Ltd

the microstructure, resistivity and transmittance of a-$Si_{1-x}Sn_x$ alloy thin films, which is crucial for the NIR brain imaging, are studied in detail.

2. Experimental Procedure

The a-$Si_{1-x}Sn_x$ alloy thin films were deposited onto Si wafers by co-sputtering. The starting materials were single crystal silicon (99.999% pure) and metallic tin (99.99% pure). The Si wafers were 1mm thick and were cleaved into 25mm × 15mm blocks to use as substrates for deposition. Sn concentration in the mixtures varied from 0% to 1%. The vacuum chamber is pumped down to a base pressure of 8.0×10^{-4} Pa. The chamber pressure was maintained at 4.3×10^{-1} Pa after introduction of Ar (99.99% pure) with a flow rate of 75 sccm. For all films discussed in this paper, Si and Sn were deposited using an RF power of 200W. Substrate temperature varied from room temperature to 300°C and the time of deposition varied from 10 minutes to 30 minutes. The films that the Sn concentration is not given in this paper have a Sn concentration of 6%. Scanning electron microscope (SEM) and transmission electron microscope (TEM) were employed to study the surface morphology of the films. The grain size of the films was characterized by X-ray diffraction (XRD). We also tested the transmission spectra and resistivity of the films to study its optical and electrical properties. Footnotes Footnotes should be avoided whenever possible. If required they should be used only for brief notes that do not fit conveniently into the text.

3. Result and Discussion

To investigate the influence of deposition time on the film quality, a-$Si_{1-x}Sn_x$ alloy thin films were deposited on Si substrates at a time range of 10-30min. Figure 1 presents the SEM images of the a-$Si_{1-x}Sn_x$ alloy thin films prepared at room temperature for 10 min (a), 20 min (b) and 30 min (c). The surface of the films is irregular at a deposition time of 10 minutes as shown in figure 1 (a), and a large number of grains having a size of about several hundred nanometers appeared. We suspect that it should be metallic Sn, and the appearance of it is due to insufficient deposition time. As the growth temperature ascends to 20 minutes, as shown in figure 2 (a), the continuity of the films were improved, irregular grains on the surface of the films did not appear, but nanoparticles around 30nm appeared. According to the following XRD test, we can infer that the nanoparticles are Sn nanocrystals. As shown in figure 1 (c), the continuity of the films is better than that of 20 minutes when the deposition time increase to 30 minutes, and the number and size of Sn nanocrystals are decreased, indicating that 30 minutes is a reasonable deposition time.

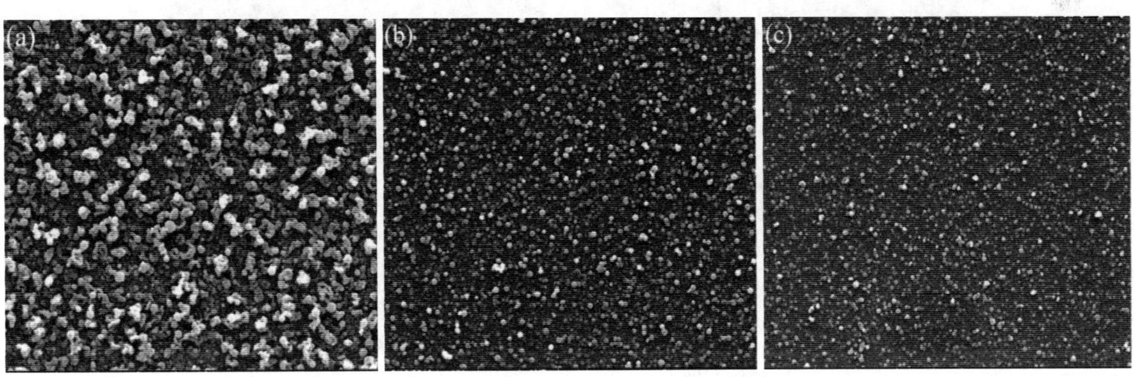

Figure 1. SEM images of a-$Si_{1-x}Sn_x$ alloy thin films at (a) 10min, (b) 20min, (c) 30min deposition time.

The growth temperature of the a-$Si_{1-x}Sn_x$ alloy thin films also has a great influence on its quality. The effect of growth temperature on the microstructure and surface morphology of the films was also studied by SEM. The SEM images of the films grown at the different temperature are shown in figure 2. The surface morphology of the films prepared at room temperature, as shown in the figure 2 (a), is not ideal, and the size of Sn grains is not uniform. As shown in figure 2 (b), when the growth temperature reaches to 150°C, the size of Sn grains becomes smaller and the distribution of Sn grains is more uniform. The improvement is the result of high solubility of Sn in silicon at high temperature. When the growth temperature reached 200°C and 250°C, the surface of the films as shown in figure 2 (c) (d) is more irregular than that prepared at room temperature. This is because during the deposition of the films, the silicon-tin binary alloy system will lower the melting point to about 200°C, and the temperature increases the solubility of Sn in silicon. However, when the deposition is completed, the solubility will start to decrease after cooling. Since the saturated solubility is reached before, as long as the temperature drops, the Sn will begin to precipitate a large amount of crystals, resulting in irregular film morphology.

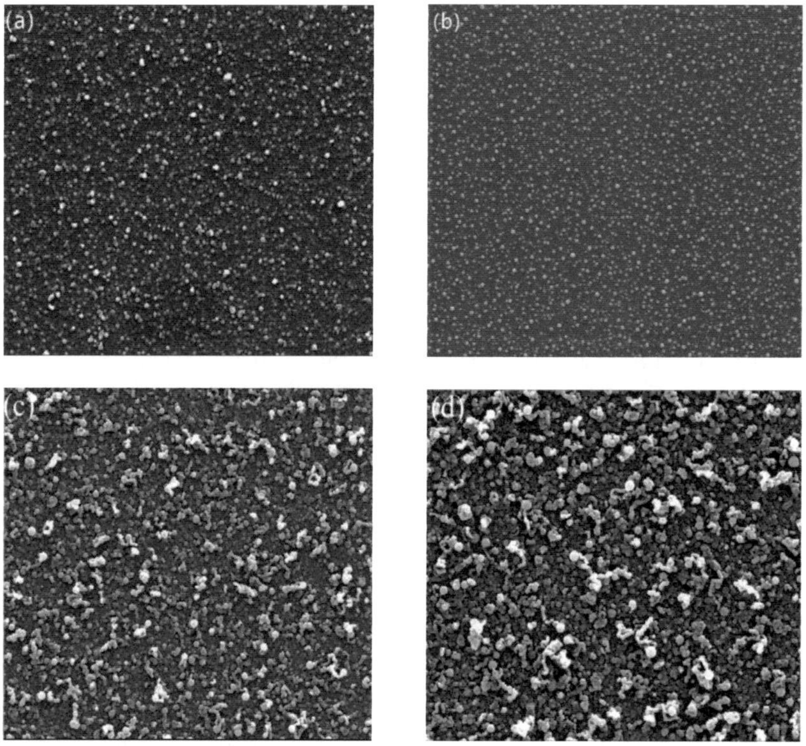

Figure 2. SEM images of a-$Si_{1-x}Sn_x$ alloy thin films grown at (a) room temperature, (b) 150°C, (c) 200°C, (d) 250°C.

Figure 3 shows the XRD patterns of the a-$Si_{1-x}Sn_x$ alloy thin films with different Sn concentration. The samples with Sn concentration lower than 1% not show any characteristic diffraction peaks for crystalline Si or crystalline Sn. As the Sn concentration in the sample increases to more than 2%, many characteristic diffraction peaks of crystalline Sn were observed. With the increase of Sn concentration, the intensity of the characteristic peak increased, indicating that more Sn nanocrystals were produced in the samples. According to Debye-Scherrer [19], we calculated that the crystallite size of Sn in the $Si_{1-x}Sn_x$ alloy thin films is about 20-40nm, which is basically consistent with the results observed in the TEM images. The XRD patterns of the a-$Si_{1-x}Sn_x$ alloy thin films not show any characteristic peaks for crystalline Si, indicating the presence of amorphous Si.

Figure 4 shows the TEM images of the a-$Si_{1-x}Sn_x$ alloy thin films grown at room temperature and 150 °C. The analysis by TEM is consistent with SEM and XRD results. It can be clearly seen in figure 4 that nanoparticles are present in the films, and it should be Sn nanocrystals by XRD pattern analysis. The size of Sn nanocrystals observed in figure 4 (a) is about 20-30nm, which is similar to that calculated by Debye-Scherrer formula. Compared with Sn nanocrystals in figure 4 (a), the number and distribution of Sn nanocrystals in figure 4 (b) become smaller and more uniform separately. The size of Sn nanocrystals is also becoming smaller, which is consistent with the results of the SEM result.

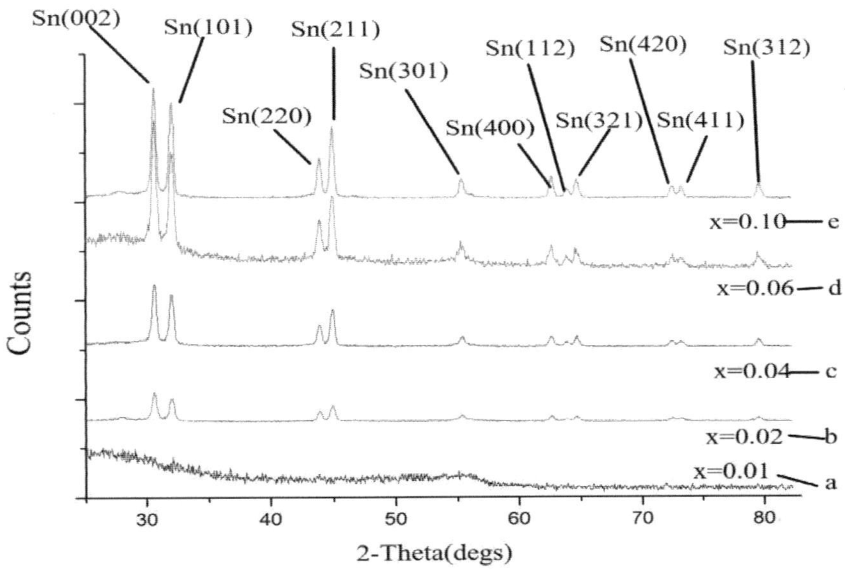

Figure 3. XRD spectra of a-$Si_{1-x}Sn_x$ alloy thin films for various values of x.f

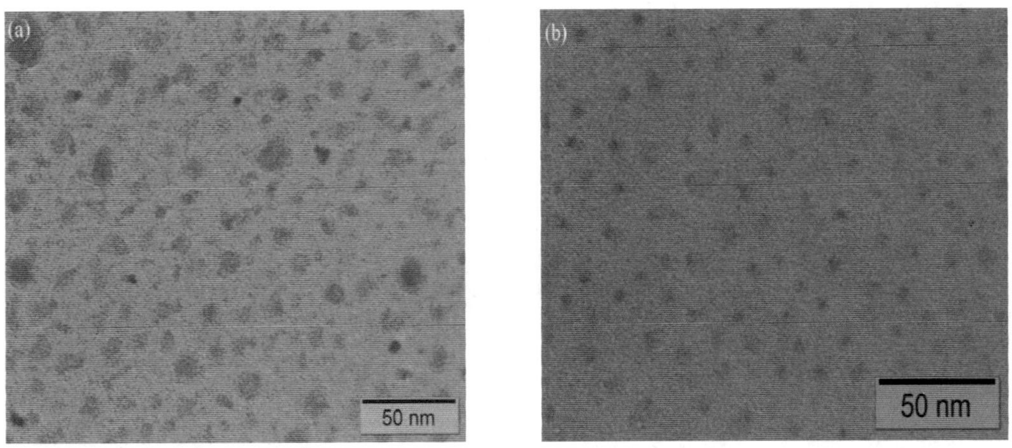

Figure 4. TEM images of a-$Si_{1-x}Sn_x$ alloy thin films grown at (a) room temperature, (b) 150°C. The insets of (a) and (b) show the related diffraction pattern.

Excellent electrical property is a prerequisite for the application of SiSn materials to near-infrared brain imaging. Therefore, it is necessary to test its resistivity. As shown in figure 5, the resistivity of the a-$Si_{1-x}Sn_x$ alloy thin films decrease about six orders when Sn concentration increases to 10%. Further, we also studied the effect of growth temperature on the resistivity of the a-$Si_{1-x}Sn_x$ alloy thin films. Figure 6 shows the change in resistivity at different growth temperatures for the films with a Sn concentration of x = 0.10. Interestingly, in the temperature range of 100-150°C, the resistivity increases with the increase of temperature. As the growth temperature is further increased, the resistivity of the films begins to increase. We suspect that this change may be caused by competition between the following two mechanisms. On the one hand, the order of the a-$Si_{1-x}Sn_x$ alloy thin films can be improved by increasing the growth temperature, and the increase in growth temperature promotes the formation of Sn nanocrystals, which reduces the dangling bonds of the lattice in the films.

These two points are the reasons for the decrease in resistivity of the a-$Si_{1-x}Sn_x$ alloy thin films. On the other hand, the increase of Sn nanocrystals consuming the Sn atoms in the Si lattice, resulting in a decrease in carrier concentration in the films, which eventually leads to an increase in resistivity.

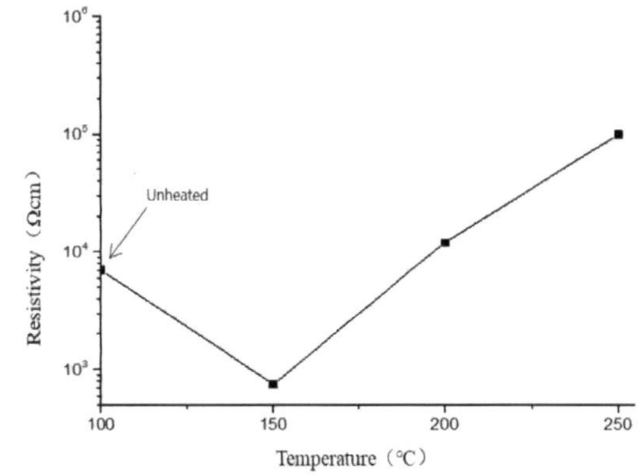

Figure 5. Resistivity change at different growth temperatures.

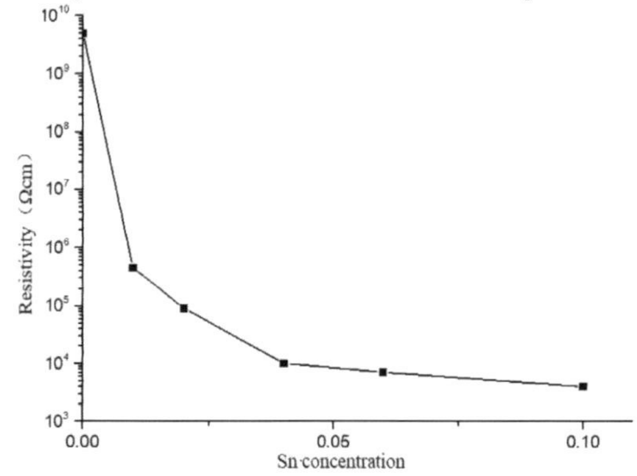

Figure 6. Resistivity of a-$Si_{1-x}Sn_x$ alloy thin films.

In addition, we also studied the ratio of photo/dark conductivity variation of the a-$Si_{1-x}Sn_x$ alloy thin films caused by different Sn concentration. The ratio of photo/dark conductivity of the films first increases and then decreases as shown in figure 7. The reason for this is because when the concentration of Sn in the films is low, the Sn atoms replace the Si atoms in the a-Si lattice to increase the bright conductance of the films, resulting in an increase in the ratio of photo/dark conductivity. When the Sn concentration of the films increases to a relatively high value, the Sn atoms cannot replace the Si atoms in the a-Si lattice, thereby providing a large amount of free electrons, resulting in an increase in the dark conductivity of the a-$Si_{1-x}Sn_x$ alloy thin films.

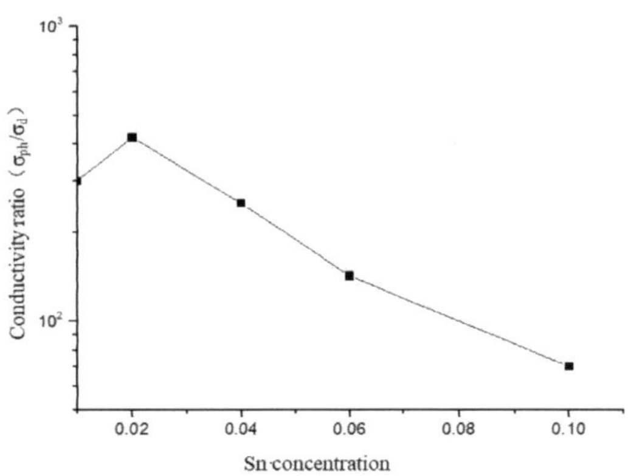

Figure 7. Ratio of photo/dark conductivity of a-$Si_{1-x}Sn_x$ alloy thin films

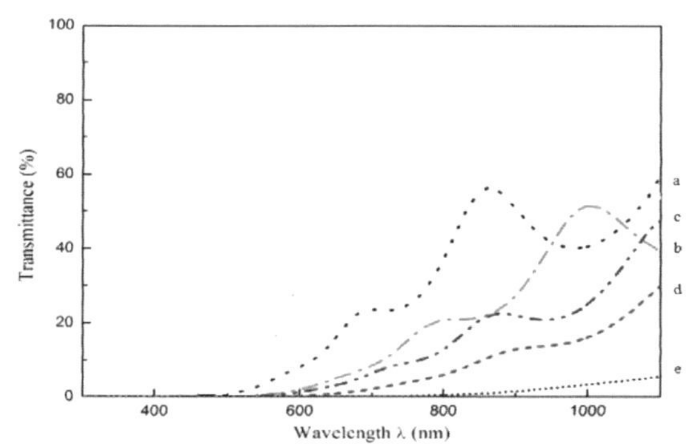

Figure 8. Transmittance spectra of a-$Si_{1-x}Sn_x$ alloy thin films for various values of x.

The transmission spectra of the a-$Si_{1-x}Sn_x$ alloy thin films were obtained by spectrophotometer. As shown in figure 8, the transmittance of the films in the visible wavelength region is very low, and tends to zero as the Sn concentration increases. The transmittance of the films in the near-infrared region decreases as the Sn concentration increases. The optical band gap of the films with different Sn concentration is shown in figure 9. The absorption wavelength corresponding to the optical band gap of the a-$Si_{1-x}Sn_x$ alloy thin films falls just in the near-infrared region, and the band gap decreases linearly with the increase of Sn concentration. The high absorbing rate will make SiSn become a promising material to be applied in near-infrared brain imaging detectors.

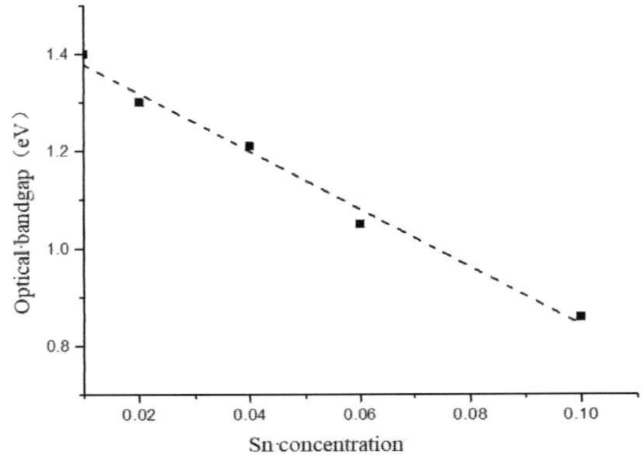

Figure 9. Optical band gap of a-$Si_{1-x}Sn_x$ alloy thin films for various values of x.

4. Conclusion

In this paper, the deposition time, growth temperature, microstructure, resistivity, and transmittance of the a-$Si_{1-x}Sn_x$ alloy thin films prepared by RF magnetron sputtering have been studied. The main conclusions of this work can be summarized as follows: (1) The reasonable deposition time and growth temperature for preparing the a-$Si_{1-x}Sn_x$ alloy thin films are 30 minutes and 150°C respectively. (2) Sn nanocaystals are dispersed in the a-$Si_{1-x}Sn_x$ alloy thin films having a size of 20-30nm. (3) Significant resistivity reduction is obtained with the increase of Sn concentration and the ratio of photo/dark conductivity increases first and then further decreases with the increase of Sn concentration. (4) The transmittance of the a-$Si_{1-x}Sn_x$ alloy thin films decreases with increasing Sn concentration in both visible wavelength and near-infrared region, and the optical band gap decreases linearly with the increase of Sn concentration.

Acknowledgments
The authors thank to the National Natural Science Foundation of China (Grant no. 51372032).

5. References

[1] E. N. Sgourou, T. Angeletos, A. Chroneos, and C. A. Londos 2017 IR studies of the oxygen and carbon precipitation processes in electron irradiated tin-doped silicon *Journal of Materials Science-Materials In Electronics* 28(14) 10298-312
[2] G. P. Ru, G. W. Wang, Y. L. Jiang, et al., 2003 Rectifying characteristics of sputter-deposited SiGe diodes *Journal of Vacuum Science and Technology B* 21(4) 1301-05
[3] M. Kurosawa, M. Kato, T. Yamaha, et al 2015 Near-infrared light absorption by polycrystalline SiSn alloys grown on insulating layers *Applied Physics Letters*106(17)
[4] A. Zaoui, M. Ferhat, M. Certier, B. Khelifa, H. Aourag 1996 Optical properties of SiSn and GeSn *Infrared Physics and Technology* 37(4) 483-8
[5] B. G. Goodyear, D. E. Brodie 1996 a-$Si_{1-x}Sn_x$:H thin films prepared by ion beam assisted deposition *Thin Solid Films* 286(1-2) 80-87
[6] J. D. Gallagher, C, Xu, L. Jiang 2013 Fundamental band gap and direct-indirect crossover in $Ge_{1-x-y}Si_xSn_y$ alloys *Applied Physics Letters* 103(20)
[7] A. M. Hussain, H.M. Fahad, N. Singh, et al. 2013 *71st* Device Research *Conf.* (Univ Notre Dame，Notre Dame，IN)
[8] A. Mohamedi, M. L. Thèye, M. Vergnat, G. Marchal, and M. Piecuch 1989 Optical studies of bonding in coevaporated amorphous silicon-tin alloys *Physical Review B* 39(6)
[9] A. S. Saidov, S. N. Usmonov, M. Kalanov, K. M. Madaminov 2010 Structure and photoelectric properties of $Si_{1-x}Sn_x$ epilayers *Technical Physics Letters* 36(9) 827-9
[10] N. Kobayashi, D. H. Zhu, H. Katsumata, et al 1996 *Proc.* of the Joint *Int.* Symposium of the '96 MRS-J Conference (Symp.N) and the *3rd* Ion Engineering *Conf.* (Chiba, Japan)

[11] J. Tolle, A. V. G. Chizmeshya, Y.-Y. Fang, and J. Kouvetakis 2006 Low temperature chemical vapor deposition of Si-based compounds via $SiH_3SiH_2SiH_3$: Metastable SiSn / GeSn / Si(100) heteroepitaxial *Applied Physics Letters* 89(23)
[12] T. D. Hatchard, J. R. Dahn 2004 Study of the electrochemical performance of sputtered $Si_{1-x}Sn_x$ films *Journal of The Electrochemical Society* 151(10) A1628-35
[13] M. Kurosawa, M. Kato, T. Yamaha, *et al.* 2014 *7th Int* Silicon-Germanium Technology and Device Meeting (Singapore)
[14] V. A. Timofeev, A. I. Nikiforrov, A. P. Kokhanenko, *et al.* 2017 Growth of epitaxial SiSn films with high Sn content for IR converters *Russian Physics Journal* 60(2) 354-9
[15] L. Y. Beaulieu, K. C. Hewitt, R. L Turner, *et al.* 2003 The electrochemical reaction of Li with amorphous Si-Sn Alloys *Journal of The Electrochemical Society* 150(2) A149-6
[16] D. Girginoudi, N. Georgoulas, and A. Thanailakis 1989 Optical and electronic properties of sputtered hydrogenated amorphous Silicon-tin alloys *Journal of Applied Physics* 66(1) 354-7
[17] H. Wang, X. Mu, J. Yang, *et al.* 2019.Brain imaging with near-infrared fluorophores *Coordination Chemistry Reviews* 380 550-71
[18] A. Abdo, M. Sahin 2007 *29th* Annual *Int Conf* of the IEEE Engineering in Medicine and Biology Society (Lyon, France)
[19] M. Shkir, S. AlFaify, V. Ganesh, I. S. Yahia 2017 Facile one pot synthesis of PbS nanosheets and their characterization *Solid State Sciences* 70 81-5

IOP Conf. Series: Materials Science and Engineering **654** (2019) 012011 doi:10.1088/1757-899X/654/1/012011

Development of Immunochromatographic Test Strip based on Quantum Dot Nanoparticles (QDs)

T Sioloetwong[1] A Kopwitthaya[4] N Sathirapongsasuti[5] K Nawattanapaiboon[1] C Kitiyakara[6] and T Srikhirin[1,2,3]

[1] School of Materials Science and Innovation, Faculty of Science, Mahidol University, 999 Phutthamonthon Sai4 Road, Salaya, Nakhon Pathom 73170, Thailand
[2] Department of Physics, Faculty of Science, Mahidol University, 272 Rama VI Road, Khawaeng Thung Phaya thai, Ratchathewi Bangkok, 10400, Thailand
[3] NANOTEC Center of Excellence, Faculty of Science, Mahidol University, 272 Rama VI Road, Khawaeng Thung Phaya thai, Ratchathewi Bangkok, 10400, Thailand
[4] National Electronics and Computer Technology Center, National Science and Technology Development Agency, 111 Phaholyothin Road, Thambon Khlong Nueng, Khlong Luang, Pathum Thani 12120, Thailand
[5] Department of Translational Medicine (Ramathibodi), Faculty of Medicine Ramathobodi Hospital, Mahidol University, 270 Rama VI Road, Khawaeng Thung Phaya thai, Ratchathewi Bangkok, 10400, Thailand
[6] Department of Medicine Ramathibodi Hospital, Faculty of Medicine Ramathobodi Hospital, Mahidol University, 270 Rama VI Road, Khawaeng Thung Phaya thai, Ratchathewi Bangkok, 10400, Thailand

toemsak.sri@mahidol.ac.th

Abstract. Water-soluble fluorescent CdSe/CdZnS quantum dots (QDs) (λ_{em}=637nm) were used as a labelling reporter for strip detection. The CdSe/CdZnS QDs was prepared by using core shell technique where CdSe QDs (λ_{em}=590nm) core was capped with CdZnS shell. CdZnS-QDs were conjugated with amine-modified DNA probe, that can bio-recognize with DNA product for use in lateral flow assays (LFAs) detection. The presence of functionalized water-dispersible QDs was confirmed by Fourier transform infra-red (FTIR) technique, where a characteristic carboxylic group was observed at 1633 cm^{-1} (-OH) and 3310 cm^{-1} (-COO). DNA probe was immobilized onto QDs by EDC/NHS pathways and confirmed by protein absorption spectra at 260 nm due to heterocyclic ring of nucleosides. QDs-DNA can be applied as labelling reporter to readout fluorescence signal on the LFAs detection.

1. Introduction

Lateral flow assay (LFAs) is a paper based technique for detecting analyte in complex mixtures without the need of costly read out systems. LFAs is widely used in many applications, such as drug abuse, ovulation monitoring, and detection of infectious organism [1-2]. The workings of LFAs are based on chromatography and immunology principles, performed on a paper based substrate. Firstly, the sample is applied to the pad where it is mixed with the labelling reporter. For a positive signal, the sample specifically interacts with the labelling reporter and moves to the test line where the sample interacts with the test molecule. Color is developed due to the presence of the labelling reporter on the test line. Conventional LFAs use gold nanoparticles (AuNPs) as labelling nanoparticles in both qualitative and semi-quantitative analysis [3].

Content from this work may be used under the terms of the Creative Commons Attribution 3.0 licence. Any further distribution of this work must maintain attribution to the author(s) and the title of the work, journal citation and DOI.
Published under licence by IOP Publishing Ltd

IOP Conf. Series: Materials Science and Engineering **654** (2019) 012011 doi:10.1088/1757-899X/654/1/012011

Yalong Bai et al. (2012) developed an assay for AFP-LFAs detection in which silica nanoparticles/quantum dots nanocomposites (SiNPs/QDs) was used to improve the assay over gold nanoparticles [4]. SiNPs/QDs showed LOD at 20 ng/ml, which was 10 times lower than AuNPs-LFAs attributed to strong fluorescence of QDs.
Nucleic acid lateral flow assay (NALFIA) based on sandwich assay of ssDNA hybridization of DNA probe and DNA product has been reported [5]. DNA probe was design to specifically interact with DNA product. The probe was immobilized on labelling reporter to readout the signal. Development of NALFIA can precisely detect various targets with a lower cost than its antibody counterpart [6]. The objective in this work is to develop a new NALFIA platform using QDs as labelling reporter for quantitative analysis with the main emphasis on the synthesis of core-shell labelling reporter and the preliminary test for its LFA function.

2. Experiment

2.1 Materials
Cadmium oxide (CdO, ≤99.99 %trace metal basis), zinc oxide (ZnO, 99.99 %trace metal basis), Sodium myristate (≥99%), Selenium (Se, 99.99 %trace metal basis), Trioctylamine (TOA, 98%), Trioctylphosphine (TOP), 1-Octadecene (ODE, technical grade 90%), Oleic acid (OA, technical grade 90%) were supplied from Sigma Aldrich. Cadmium nitrate tetrahydrate ($CdNO_3$•$4H_2O$),(99 %metals basis), Sulfur (S, 99.999 %metals basis, 3-mercaptopropionic), MPA, (99%), Tetramethylammonium hydroxide (TMAHP, 1.0M) were purchased from Alfa Aesar.

2.2 Synthesis of core CdSe QDs [7]
174 mg of Cadmium myristate and 12 mg of selenium were dispersed in 16 ml of 1-octadecene in 3 necked round bottom flask equipped with a condenser. The reaction was stirred vigorously under vacuum for 1 hour. After which, the vacuum was removed. To prevent the oxidation reaction, the reaction was kept under a nitrogen atmosphere using a balloon filled with nitrogen gas. The reaction was heated up to 240^0C by using aluminium heating block and held for 40 minutes. 1 mL of oleic acid, capping agent, was immediately added to the reaction to activate the formation of nanoparticle. The reaction was held at 240^0C for 2 minutes before air cooled down to room temperature. EtOH was added to the reaction when phase separation occurred. The supernatant was removed and the bottoms portion was kept in 10 mL of chloroform (CHCl3).

2.3 Synthesis of core-shell CdSe/CdZnS CSQDs [7]
400 μL of $Zn(oleate)_2$, 200 μL of $Cd(oleate)_2$, 8 mL of TOA, and 2 mL of QDs were added into a round bottom flask. The mixture was heated up to 70^0C and held for 1 hour. The reaction was subsequently heated up to 300^0C under N_2 atmosphere, and mixtures of TOP-S and TOA with the volume ratios of 1:10 were added, and held for 2 hours . After the reaction turned into darker color, the CSQDs was cooled down to room temperature and washed with the same step as the core CdSe QDs.

2.4 Carboxylic acid modification of CSQDs surface [8]
20 mL of binary solvent was prepared by mixing MeOH and 1,4-dioxane at 1:1 ratio. pH of binary mixture was adjusted to pH 13 -14 by using 0.5 M TMAHP buffer in order to convert the charge of 3-MPA into anionic. 88 μL of 3-MPA was added to the binary mixture and was heated up to 70^0C with vigorously stirring. 200 μL of 1 μM CSQDs was added to promote the sulfur ion to form disulfide bond. The mixture was held for 15 minutes in order to promote the intermolecular interaction of disulfide bond on CSQDs surface. The mixture was held at 70^0C under vigorous stirring for 15 min, and 3 mL of EtOAc was added into the reaction to promote the precipitation of particle. The mixture was centrifuged at 4000 rpm for 40 min in order to remove free MPA, and yielding CSQDs-COOH. CSQDs-COOH was washed by centrifugation and dispersed in 1 mL of DI water.

2.5 QDs-DNA probe immobilization [9]
1 μM of CSQDs-MPA solution was mixed with 40 μL of 10 mM EDC and NHS with the volume ratios at 1:1. The solution was shaken for 10 minutes before the addition of 50 μL of 40 μM DNA probe .After DNA loading, the CSQDs-DNA was shaken continuously for another 10 hours. QDs-DNA probe was centrifuged for supernatant depletion, re-dispersed in borate solution, and stocked in refrigerator before use .

2.6 Assembly strip of biosensor
The fabrication strip consisted of three major parts: (i) sample pad, (ii) nitrocellulose pad, and (iii) absorbent pad, which were all pasted on PVC sheet. All segments overlapped by around 4 mm. The assembled strip was cut into 4 mm x 6 cm rectangular shape. The assembled strips were kept at 4 ^{0}C in a refrigerator.

3. Results and discussions

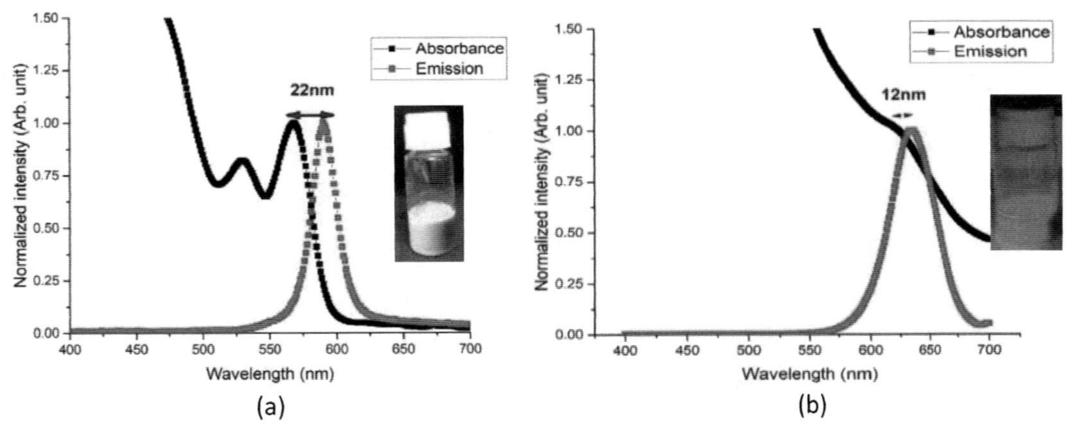

Figure 1. Normalized intensity of optical properties: (a) core CdSeQDs and (b) CdSe/CdZnS CSQDs

Figure 1 showed the stroke shift of maximum absorption wavelength and emission spectrum of both (a) core CdSeQDs and (b) Core-shell CdSe/CdZnS CSQDs. The maximum emission wavelength of core CdSeQDs was located at 590 nm and was related to the visible color emission in yellow. FWHM of emission wavelength of core-shell CdSe/CdZnS CSQDs was 22 nm .The size approximation was calculated by fitting function that provided by :

$$D\ (nm) = 1.6122\times10^{-9}(\lambda^4) - 2.6575\times10^{-6}(\lambda^3) + 1.6242\times10^{-3}(\lambda^2) - 0.4277 \quad (1)$$

where D nm (is diameter of nanocrystal QDs, and λ nm is first excitonic absorbance peak). The diameter of CdSe QDs was calculated to be 3.5 nm, 3.3 nm, and 3.4 nm. CSQDs showed the stroke shift phenomena as in the case of core CdSe QDs that was related to its fluorescence property .The fluorescence emission wavelength were was compared. The red shift occurred from 623 nm of core QDs to 635 nm of CSQDs, based on quantum confinement effect as the size increased Bohr's radius [10].

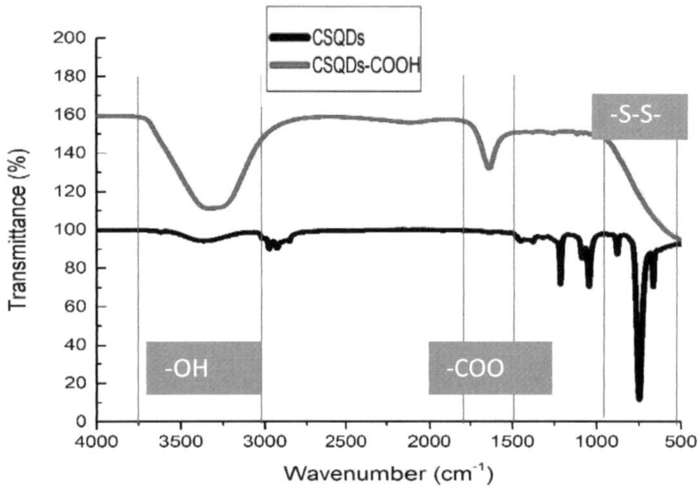

Figure 2. FTIR spectra of CSQDs and CSQDs-COOH

FTIR spectra in figure 2 showed the characteristic peaks of surface functional groups. Absorption bands of CSQDs-MPA spectra located at 2950-3300 cm^{-1} for -OH stretching and at 1650-1700 cm^{-1} for -COO stretching of carboxylic group .The disulfide stretching band was located at 500-600 cm^{-1}, which coordinated with Cd ion on surface of CSQDs and thiol group .

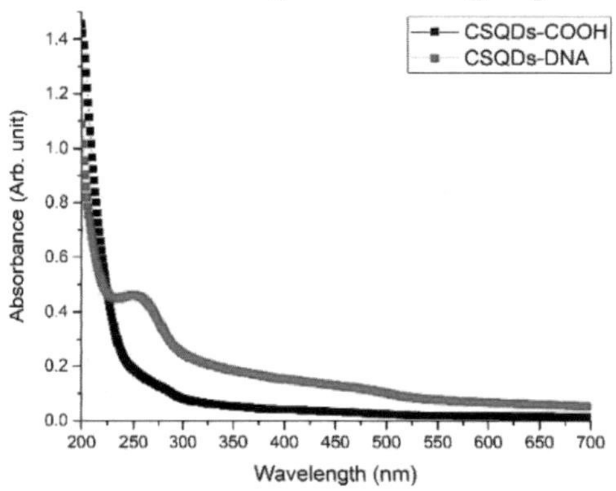

Figure 3. Absorbance spectrum of modified QDs and QDs-DNA

Figure 3 showed absorption peak around 260 nm that was due to protein adsorption on QDs surface from heterocyclic of nucleosides when compared with neat modified QDs.

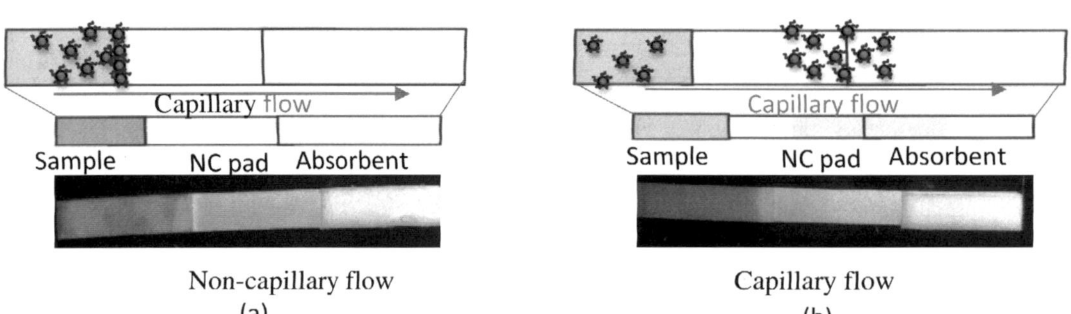

Figure 4. Capillary flow behaviour of QDs-labelled strip biosensor: (a) Non-capillary flow and (b) Capillary flow.

Capillary flow behavior of QDs-labelled strip biosensor was showed in figure 4. The non-capillary flow showed adherence of QDs on the sample pad which arose from the particle solubility in the running buffer. In a good solubility buffer, QDs could flow from the sample pad to absorbent pad. All results showed red emission color of QDs under UV excitation.

4. Conclusion

QDs were successfully synthesized and immobilized with DNA probe for use as labelling reporters. Using the current model, QDs-DNA could be applied to lateral flow assay for DNA detection and fluorescent signal can be readout under UV excitation.

Acknowledgements

This research was mainly supported by TGIST scholarship from NSTDA and CIF scholarship. The authors would like to thank Dr. Atcha Kobwitthaya (NECTEC), Dr.Nuankanya Sathirapongsasuti, and Asst. Prof. Dr.ToemsakSrikhirin for their collaborations with the project.

5. References

[1] Koczula M K., and Gallotta A 2016 Lateral Flow Assays *Essays Biochem.* 60 111-120
[2] Bahadir B E., and Sezgintürk K M 2016 Lateral Flow Assays: Principles, Designs and Labels *Trends Analyt Chem.* **82** 286-306
[3]Gong X., and Chang. J 2017 A Review for Fluorescent Lateral Flow Immunochromatographic Strip *J. Mater. Chem.* B, **5** 5079-5091
[4] Bai Y., Tian C., Wei X., Wang Y., Wanga D., and Shi X 2012 A Sensitive Lateral Flow Test Strip Based on Silica Nanoparticle/CdTe Quantum Dot Composite Reporter Probe *RSC Advances* **2** 1778-1781
[5]M. Blažková M., Koets M., Wich ers H J., Amerongen van A., Fukal L and Rauch P 2009 Nucleic Acid Lateral Flow Immunoassay for the Detection of Pathogenic Bacteria from *Food Czech J. Food Sci.* **27** S350-3
[6] Javani A., Javadi-Zarnaghi., and Rasaee J M 2017 Development of A Colorimetric Nucleic Acid- Based- Lateral Flow Assay with Non-biotinylated Capture DNA *Appl. Biol. Chem.* **60** 637-645
[7] Carion O., Mahler B., Pons T., and Dubertret B 2007 Synthesis, Encapsulation, Purification and Coupling of Single Quantum Dots in Phospholipid Micelles for Their Use in Cellular and in Vivo Imaging. *Nat. protoc.* **2**, 2383-2390.
[8]Vladimir V. Breus V., Pietuch A., Tarantola M., Basché T., and Janshoff A., 2015 The Effect of Surface Charge on Nonspecific Uptake and Cytotoxicity of CdSe/ZnS Core/shell Quantum Dots. *Beilstein J. Nanotechnol.* **6**, 281–292.
[9] Yang J., Songb J T., Hua Wei., Huanga Yan., Denga L J., Houa H. N., and Liu B 2017 Fabrication of Fluorescence Enhancement of Quantum Dots on A Gold Colloid Formed Film for Oligonucleotide DNA Detection. *Anal. Methods* **9**, 434–44
[10]Yu WW, Qu L., Guo W., and Peng X 2003 Experimental Determination of The Extinction Coefficient of CdTe, CdSe, and CdS Nanocrystals *Chem. Mater.* **15** 2854-2860.

IOP Conf. Series: Materials Science and Engineering **654** (2019) 012012 doi:10.1088/1757-899X/654/1/012012

Chip Geometry Modelling of 2-Dimension Ultrasonic Vibration Assisted Machining (2D UVAM) on Micromilling Machine With End Mill

Gandjar Kiswanto[1], Y R Johan[1], Poly[1] and T. J. Ko[2]

[1]Department of Mechanical Engineering, Universitas Indonesia, Depok 16424, Indonesia
[2]School of Mechanical Engineering, Yeungnam University, Republic of Korea

gandjar_kiswanto@eng.ui.ac.id

Abstract. The needs of miniaturized products have increased a lot in this ever-changing world. This makes the micromanufacturing technologies develop fast in order to keep up with this higher needs and to meet the required quality of a product. One of the developed technologies is ultrasonic vibration assisted machining (UVAM). UVAM is different than conventional machining because of the way the cutting tool move relative to the workpiece. This different cutting phenomenon produces a different chip geometry than the conventional machining. The purpose of this paper is to give an understanding through chip geometry models about how UVAM can be a better cutting method rather than conventional milling. MATLAB is used in order to do the modelling of theoretical chip geometry. The approach used in this study is by calculating where the cutting tool edge is in a given unit of time before and after the ultrasonic vibration is induced to the workpiece in X-Y dimension. A characteristic in the cutting edge called the bottom cutting edge angle is also considered. By comparing the two chip geometries, some benefits that UVAM gives are explained.

1. Introduction

The escalating use of portable electronic devices makes manufacturing micro parts with high precision essential. Not only electronic industry, many industries that require hard and brittle materials makes the the development of manufacturing technologies needed to satisfy those demands increases. A manufacturing technology neededed basically require these following statements:

- The technology needs to produce better quality products.
- The technology must cheap enough to suppress manufacturing cost.

In 1979, the new technology for machining was developed. It is reported that a steel can be machined with diamond tool by applying ultrasonic vibration. The result was not achieved to ultra-precision level but then, in 1991 Moriwaki and Shamoto [1] developed ultrasonic vibration tool for ultra-precision diamond turning with a fulfilling result.

Ultrasonic vibration assisted machining is a method used in cutting where vibration with an ultrasonic frequency (higher than 20 kHz) and certain amplitudes are applied either to cutting tool or to the workpiece. Because there is vibration, the relative motion between cutting tool and the workpiece is different compared to the conventional machiing.

The advantages of UVAM compared to conventional machining have been studied experimentally [2, 3, 4] and through simulations [5,6]. Rasidi et al [6] have presented the advantages of UVAM by comparing the chip thickness. There is also a study about the cutter-workpiece engagement [7]. However, there are still little study about the cutter-workpiece engage-disengagement and the chip geometries for UVAM. Understanding more about these can analyze further about what the advantages of UVAM are. Chip geometries are generated by calculating where the cutting edge is.

Content from this work may be used under the terms of the Creative Commons Attribution 3.0 licence. Any further distribution of this work must maintain attribution to the author(s) and the title of the work, journal citation and DOI.
Published under licence by IOP Publishing Ltd

IOP Conf. Series: Materials Science and Engineering **654** (2019) 012012 doi:10.1088/1757-899X/654/1/012012

This paper takes into account some characteristics of the cutting tool such as bottom cutting edge to show deeper understanding about the advantages of UVAM.

2. Calculating the cutting edge position

There are several mathematical model used in order to generate the chip geometry. Before looking into them, there are several given parameters used in order to create this modelling simulation. There are cutting parameters, e.g., spindle speed, feed rate, feed per tooth, and depth of cut. Another one is, vibration parameters, e.g., ultrasonic frequency, amplitude (less than 6 μm [8]). Also, since this is a micro-scale machining, the precense of the tool edge radius, which is comparable in size to the uncut chip thickness itself, introduces a minimum uncut chip thickness (MUCT) [9,10].

2.1. Conventional machining

Chip geometry is generated because there is angular displacement caused by spindle speed and also linear displacement caused by feed rate as seen from figure 1 and figure 2.

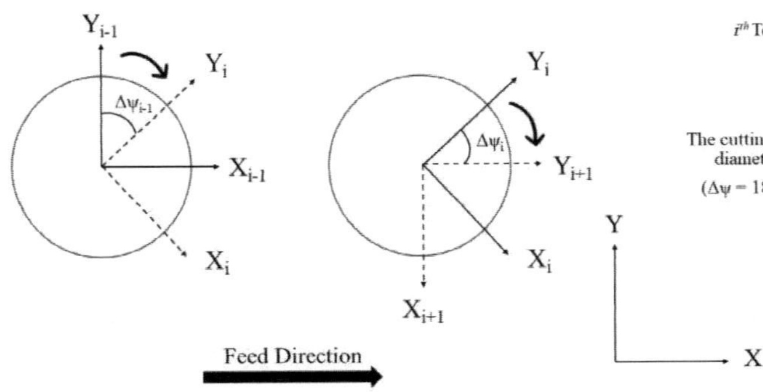

Figure 1. Angular displacement of the cutting edge.

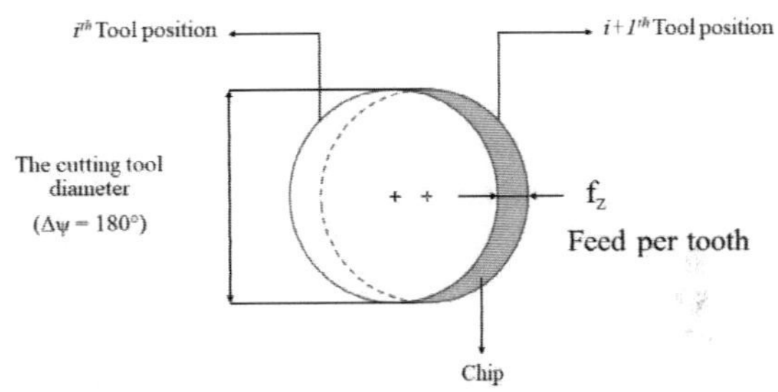

Figure 2. Feed motion linearly.

For calculating the feed per tooth, equation (1) below is used.

$$fz = \frac{v}{N \times n} \tag{1}$$

Where v is feed rate (mm/min), N is spindle speed (RPM), and n is number of flute. While the cutting tool edge position can be calculated as equation (2) and (3) below.

$$X = R \sin \Delta\psi \tag{2}$$

$$Y = R \cos \Delta\psi \tag{3}$$

Where X is cutting tool edge position on X axis with reference *to tool center coordinate system* (mm), Y cutting tool edge position on Y axis with reference *to tool center coordinate system* (mm), R is the cutting tool radius (mm), $\Delta\psi$ = cutting tool edge rotation angle around Z axis (°).

According to Kouravand and Imani's work [11], tool center coordinate system is a coordinate system defined to express the center position of the cutting tool on the center of the channel. The origin of the coordinate system is located under the tool axis. In this study, feed direction is along the X axis.

2.2. UVAM

On UVAM system, the cutting tool moves towards the workpiece then makes a contact with it until the given positive amplitude (engage), then the cutting tool will move away from the workpiece until it fully loses contact with the workpiece (disengage). Cerniway [12] said that the displacement of cutting tool caused by vibration in a given time domain can be calculated with equation (4) and (5).

$$Dx(t) = A \cos(\omega t) + fz \tag{4}$$

$$Dy(t) = A \sin(\omega t) \tag{5}$$

Where A is amplitude (mm), f is frequency (Hz), t is time and $\omega = 2 \times \pi \times f$.

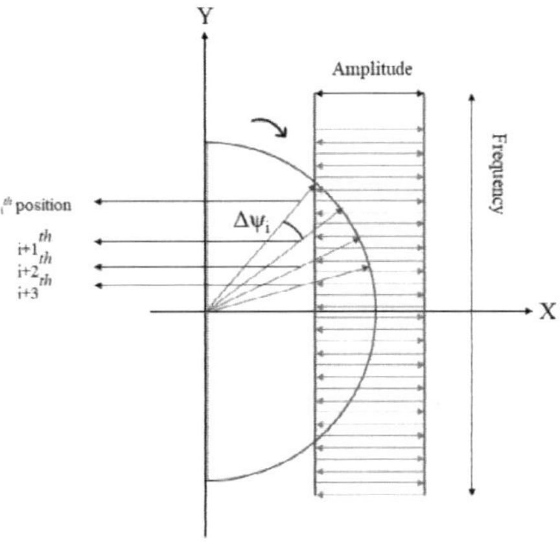

Figure 3. Schematic of vibration cutting

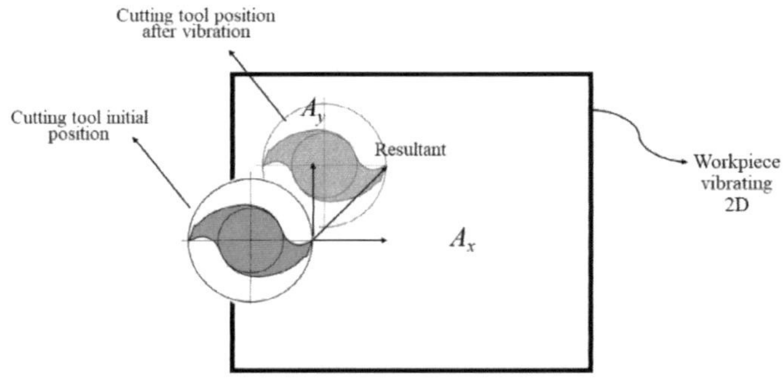

Figure 4. Resultant of 2D vibration

From figure 3 above, it shows how the relationship between amplitude and frequency of UVAM system. The illustration only shows 1D vibration cutting for the purpose of simplification and to give clearer look. However in figure 4, it shows when the amplitudes are produced on each axis. In this study, the phase different, or the time difference when voltage is generated to each axis' actuator, is 0. This makes the cutting tool moves linearly in a diagonal way instead of elliptical motion.

2.3. *Bottom cutting edge feature*

Kouravand and Imani [11] introduces bottom cutting edge angle in their machined surface roughness modelling.

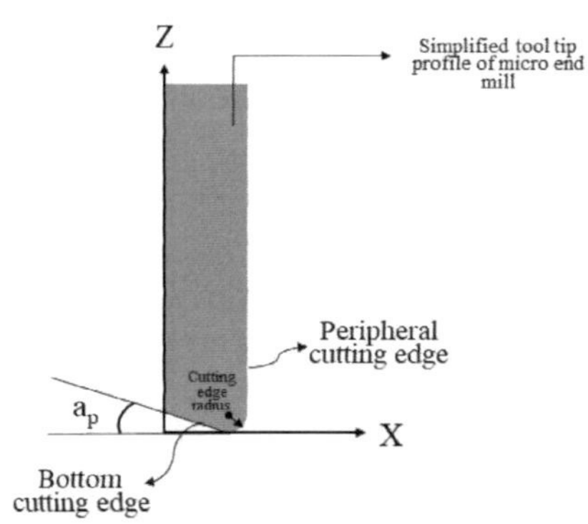

Figure 5. Cutting edge of end-mill

When there is a prescense of bottom cutting edge angle, as illustrated on figure 5, the axial depth of cut will varry. In this paper, by analytical approach, the varrying can be calcculated with equation (6) and equation (7).

$$Z_i = Doc \quad (6)$$

$$Z_{i-1} = Doc - \tan(a_p \times fz) \quad (7)$$

3. Chip geometries

The chip geometries that are generated can be seen in figure 6 and figure 7 below. The cutting parameters used are axial depth of cut 10 µm, feed per tooth 0.012 mm and spindle speed 50000 RPM. The vibration parameters are amplitude 1 µm and frequency of 24 kHz. Cutting tool used is 2 flute end-mill DIXI 7242 with diameter of 1 mm, and bottom cutting edge angle 3.4393°.

In figure 6, the presence of bottom cutting edge angle creates height difference of the material removed or the chip. For example, in figure 6(a), the red line has shorter height than the blue line. This means the machined surface will have a natural roughness caused by this cutting edge feature.

IOP Conf. Series: Materials Science and Engineering **654** (2019) 012012 doi:10.1088/1757-899X/654/1/012012

When removing a material in conventional machining, the right combination of cutting parameters needs to be applied to create better surface quality. Many studies have suggest that lower feed rate can produce better surface quality. This lower feed rates creates less height different in the chip. However, making the feed rate smaller does not always guarantee the good surface quality, even with the right range of spindle speed. This is because there is a minimum chip thickness value that needs to be satisfied in order for chip to be developed. Even when this smaller feed rate is big enough to meet the minimum chip thickness value, the machining time would take longer and creates inefficiency.

In figure 7, the engage and disengage phenomenon can be looked better. When the amplitude is maximum, the cutting edge will have an offset position when compared to the conventional milling. This offset position can act as an additional feed per tooth, making the natural roughness caused by bottom cutting edge angle being cutted or machined, without the needs to reconsider smaller feed per tooth and longer machining time.

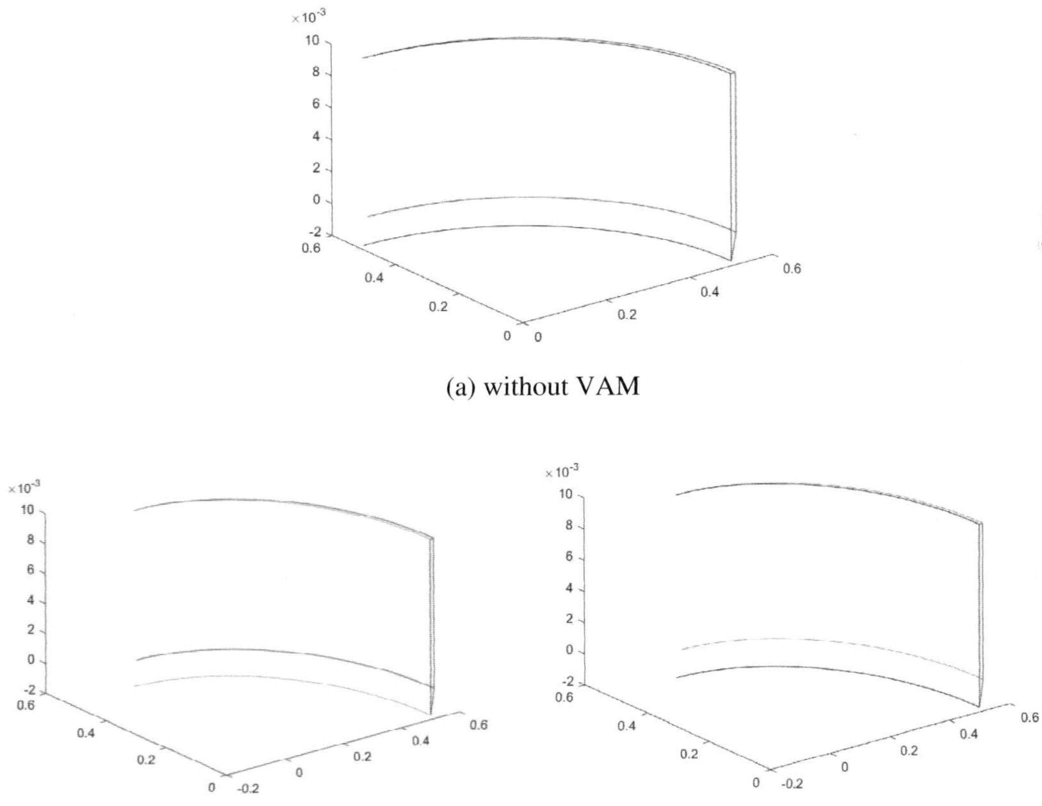

(a) without VAM

(b) with 1D VAM (c) with 2D VAM

Figure 6. Isometric view of chip geometry

Chips can have many shapes [13]. The desired chip is usually moderate, not too long and also not discontinuous. Longer chip can cause built-up edge (BUE) on the rake tool face. BUE will create a rake wear [14]. Hence, thinner chip is preferred. UVAM creates thinner chip so it gives higher tool life compared to conventional milling.

In figure 7(c), interestingly, although the simulation only used 1D VAM, the elliptical motion is generated. This makes sense, because cutting tool is also rotating while the exited vibration is only through the X axis, along the feed direction. On other hand the 2D VAM with no phase difference creates the resultant motion linearly in diagonal way. It can be possible that 1D VAM along the feed direction can gives better overall machining qualities compared to the non elliptical 2D VAM.

IOP Conf. Series: Materials Science and Engineering **654** (2019) 012012 doi:10.1088/1757-899X/654/1/012012

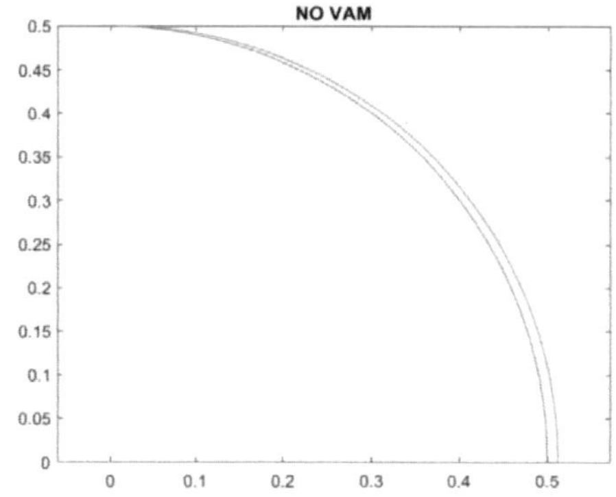

(a) without VAM

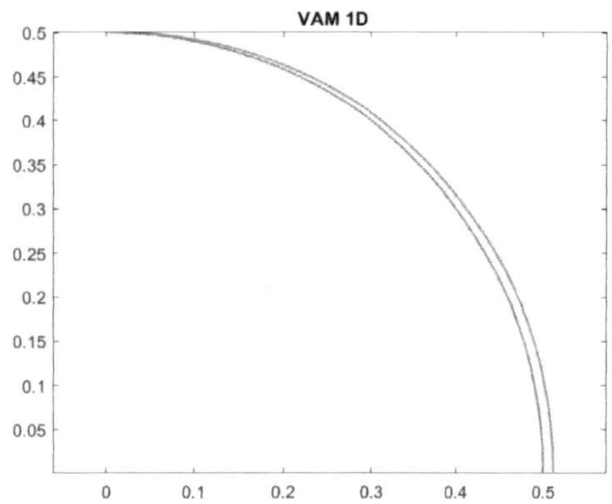

(b) with 1D VAM

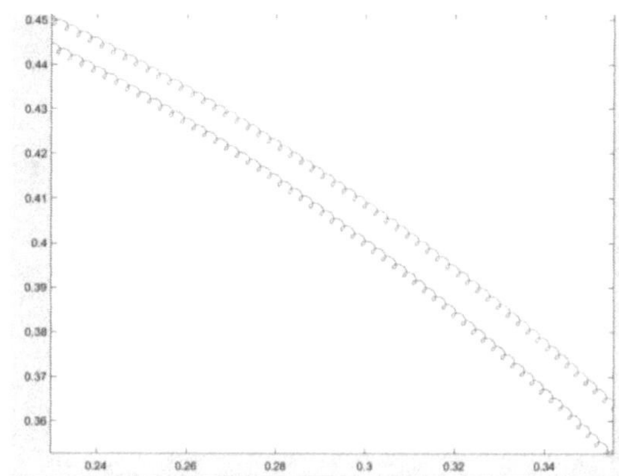

(c) Magnified 1D VAM

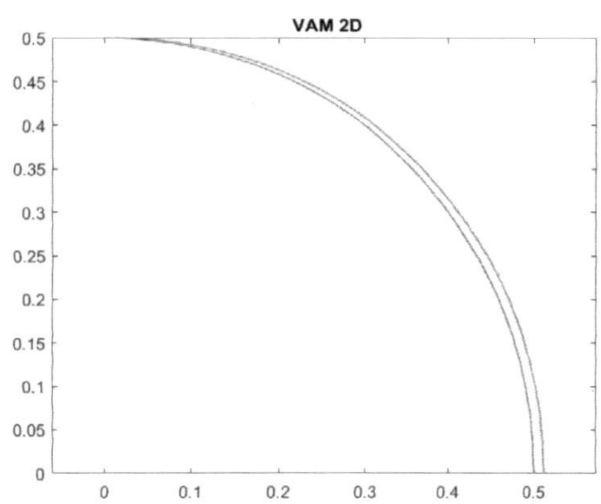

(d) with 2D VAM

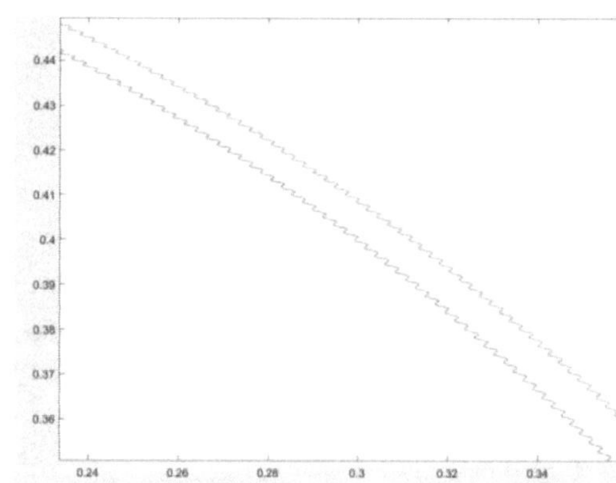

(e) Magnified 2D VAM

Figure 7. Top view of chip geometry

In the given amount of time, both conventional and UVAM remove the same amount of material. However, in UVAM, there is a moment where disengage between cuttting tool and the workpiece happen which makes the occurrence of removing this amount of material to be divided by several parts [12]. This is why the cutting force could be reduced using UVAM.

4. Conclusion and suggestion

This paper gives an understanding about why UVAM can give some advantages rather than conventioal machining using the model of each chip geometries. The reasons are as follows:

- The desired chip shape in machining is moderate, not too long and not discontinuous. UVAM makes the removal of material more efficient by making the chip shape thinner than conventional machining.
- The engage and disengage cutting tool and the workpiece by UVAM causes the cutting force needed is going to be smaller compraed to the conventional machining.
- The extra feed per tooth caused by the positive amplitude in vibration creates a greater displacement of the cutting tool to the workpiece, hence making the natural roughness of the workpiece getting machined and generating better surface quality.
- Thinner chips are produced instead of thick chip means rake wear caused by built-up edge will not happened as quickly as conventional machining so it increases tool life.

The elliptical motion in 2D UVAM is usually more commonly used. Understanding more about the phase different to create the chip geometry needs to be studied in depth.

Acknowledgments
Authors acknowledges the financial support from PIT 9 Research Grant 2019-Universitas Indonesia with grant number NKB-0057/UN2.R3.1/HKP.05.00/2019.

5. References

[1]Moriwaki, T., & Shamoto, E. (1991). *Ultraprecision Diamond Turning of Stainless Steel by Applying Ultrasonic Vibration. CIRP Annals, 40(1), 559–562.* doi:10.1016/s00078506(07)62053-8

[2]D.E. Brehl, T.A. Dow, Review of vibration-assisted machining, Precis. Eng. 32 (2008) 153172

[3]El-Taybany, Y., M. Hossam, and H. El-Hofy, *Experimental Investigation of Ultrasonic-Assisted Milling of Soda Glass Using Factorial Design of Experiments.* Procedia CIRP, 2017. **58**(Supplement C): p. 381-386 DOI: https://doi.org/10.1016/j.procir.2017.03.238

[4]Kurniawan, R., Kumaran, S. T., Ali, S., Nurcahyaningsih, D. A., Kiswanto, G., & Ko, T. J.(2018). *Experimental and analytical study of ultrasonic elliptical vibration cutting on AISI1045 for sustainable machining of round-shaped microgroove pattern. The International Journal of Advanced Manufacturing Technology, 98(5-8), 2031–2055.* doi:10.1007/s00170018-2359-1

[5] Ding, H., Chen, S.-J., & Cheng, K. (2010). *Dynamic surface generation modeling of two-dimensional vibration-assisted micro-end-milling. The International Journal of Advanced Manufacturing Technology, 53(9-12), 1075–1079.*doi:10.1007/s00170-010-2903-0

[6]I Rasidi, I., Rafai, N. H., Rahim, E. A., Kamaruddin, S. A., Ding, H., & Cheng, K. (2015). *An investigation of cutting mechanics in 2 dimensional ultrasonic vibration assisted milling toward chip thickness and chip formation. IOP Conference Series: Materials Science and Engineering, 100, 012057.* doi:10.1088/1757-899x/100/1/012057

[7]Kiswanto, G., Hendriko, H., & Duc, E. (2015). *A hybrid analytical- and discrete-based methodology for determining cutter-workpiece engagement in five-axis milling. The International Journal of Advanced Manufacturing Technology, 80(9-12), 2083-2096.* doi:10.1007/s00170-015-7156-5

[8]Kumar, M. N., Subbu, S. K., Krishna, P. V., & Venugopal, A. (2014). *Vibration Assisted Conventional and Advanced Machining: A Review. Procedia Engineering, 97, 1577–1586.*doi:10.1016/j.proeng.2014.12.441

[9]Malekian, M., Mostofa, M. G., Park, S. S., & Jun, M. B. G. (2012). *Modeling of minimum uncut chip thickness in micro machining of aluminum. Journal of Materials Processing Technology, 212(3), 553–559.*doi:10.1016/j.jmatprotec.2011.05.022

[10]De Oliveira, F. B., Rodrigues, A. R., Coelho, R. T., & de Souza, A. F. (2015). *Size effect and minimum chip thickness in micromilling. International Journal of Machine Tools and Manufacture, 89, 39–54.*doi:10.1016/j.ijmachtools.2014.11.001

[11]Shahriar Kouravand & Behnam Moetakef Imani (2014) Developing A Surface Roughness Model for End-Milling of Micro-Channel, Machining Science and Technology: An International Journal, 18:2, 299-321, DOI: 10.1080/10910344.2014.897846

[12]Cerniway, M.A. (2001). Elliptical Diamond Milling: Kinematics, Force and Tool Wear.Master's Thesis. North Carolina State University

[13]Grzesik, W. (2017). *Chip Formation and Control. Advanced Machining Processes of Metallic Materials, 113–146.* doi:10.1016/b978-0-444-63711-6.00007-7

[14]Grzesik, W. (2017). *Tool Wear and Damage. Advanced Machining Processes of Metallic Materials, 215–239.* doi:10.1016/b978-0-444-63711-6.00012-0

IOP Conf. Series: Materials Science and Engineering **654** (2019) 012013 doi:10.1088/1757-899X/654/1/012013

The Effects of Cutting Parameters to the Surface Roughness in Low Speed Machining of Micro-milling of Titanium Alloy Ti-6Al-4V

Gandjar Kiswanto[1], A Mandala[1], M Azmi[1] and T. J. Ko[2]

[1] Department of Mechanical Engineering, Universitas Indonesia, Indonesia
[2] School of Mechanical Engineering, Yeungnam University, Republic of Korea

gandjar_kiswanto@eng.ui.ac.id

Abstract. The demand for micro-scale products is increasing rapidly in various fields of industries such as electronics, bio-medical, optical industry, and so on. Titanium alloys especially Ti-6Al-4V is one of the commonly used in bio-medical industries because of its biocompability properties. However, poor surface quality in terms of surface roughness commonly occurs because of unappropriate cutting parameters to machine this hard to cut material. This study aims to investigate the influencing machining parameters to produce micro-products with a low level of surface roughness in Titanium Alloy (Ti-6Al-4V) material using a miniaturized micro-milling machine. Experiments carried out by micromilling process with variations in low rpm spindle speed and feed rate with a constant depth of cut using a carbide cutting tool of with a diameter of 1 mm. The machining results in the form of a 4 mm slot with a depth of 10 μm, which then measures its surface roughness. It was found that as the feed rate increases, the surface roughness also accordingly increases. On the other hand, the surface roughness decreases as the spindle speed increases.

1. Introduction

The increasing of demand of micro products in recent decades are getting higher attention for bio medical researches. Material upgrades are one of the optimizations of the product that can make it stronger and more efficient. The characteristics of the micro products are including functional sizes in the range of 1 to 100 μm, high levels of precision, generally less than 1 μm of tolerances, very small surface roughness with less than 0.5 μm of R_a , a complex shape and design, and uses heat resistance materials such as ceramics, titanium alloys etc [1].

Titanium alloys are commonly used for bio-medical micro-product for its bio-compatibility, corrosion resistance, and a lightweight yet strong material. Implants, and dental products such as crowns, dentures, etc. are using titanium alloys [2]. However, poor machinability of titanium alloys is the main problem of machining this hard to cut material. Thepsonthi et. al. [3] stated that rapid tool wear and severe burr formation is one of the difficulties for machining of titanium alloys. Micro-milling is one of the choices for 3D micro-manufacturing processes, especially for complex micro product that need high accuracy and high surface quality without further finishing process such as grinding and polishing. Despite of requiring a relatively longer, micro-milling has the ability to make products from various materials with complex shape.

To achieve the best surface quality, an optimal cutting parameter have to be obtained. In micro-milling, selection the right : cutting tool , feed rate, spindle speed, and depth of cut is the most

Content from this work may be used under the terms of the Creative Commons Attribution 3.0 licence. Any further distribution of this work must maintain attribution to the author(s) and the title of the work, journal citation and DOI.
Published under licence by IOP Publishing Ltd

important to do since the quality of the process totally rely on those parameters. There are several studies that present the challenges of selecting cutting parameters in the micro-milling process. Sharp cutting edge will cause chatter vibration at a relatively small depth of cut where the cutting depth of the micro-milling process is generally very small, less than 500 μm [4]. Bandapalli et al [5] already investigated the tool wear on various coated cutting tools for titanium alloy high speed micro-milling. It is found that uncoated carbide cutting tool performed better than any coated carbide tools in titanium alloy high speed micro-milling process. Wang et al [6] investigated the errors in micro-milling. It is found that increased surface roughness is directly proportional to the increase in tool diameter and spindle speed. In addition, certain range of spindle speed can cause vibration and affect surface roughness. To improve the quality of the workpiece surface roughness, the stiffness of the tool structure must be increased. This can be done by reducing the distance of the cutting tool with the collet.

However, from literature review, it is found that most researches in machining titanium alloy Ti-6Al-4V carried out high speed machining at the range of 10,000 rpm to 120,000 rpm [3][5][7][8]. The main reason is that high speed machining produces better surface quality with high material removal rate (MRR). On the other hand, low speed machining still has a potential of offering high quality machined surface with an appropriate low budget cutting tool and low budget milling machine that have an ability of producing spindle speed only up to 10,000 rpm. Bach et al [9] investigated the effect of cutting parameters to surface quality in a conventional titanium alloy Ti-6Al-4V low speed milling. It was stated that low speed milling offers high stability that would have ability to machine in a high depth of cut with high surface quality. Moreover, lower feed rate generally produced smoother surface [10]. Thus, this study aims to investigate the influence of the cutting parameters to the surface roughness of low speed Ti-6Al-4V micro-milling.

2. Experimental Setup

The Titanium alloy Ti-6Al-4V micro-milling experiment was conducted using a miniaturized 5-axis micro-milling machine as shown in Figure 1. The machine has three linear axes of XYZ and two rotational axes of A & C but only the linear axes are used in this experiment. The movement of the axes are powered by Suruga Seiki stepper motors with 1 μm of motion resolution for each axis which the speed, direction, and distance are controlled by DS102 motor controller from Suruga Seiki.

(a) **(b)**

Figure 1. (a) Miniaturized 5-axis micro-milling machine (b) Cutting process.

The workpiece material that is used was Ti-6Al-4Al with a hardness of 44 HRC with the dimension of 10 mm × 10 mm × 4.82 mm. DIXI 7242 uncoated carbide cutting tool that has 2 flutes as shown in Figure 2 and has a specification as shown in Table 1 was used in the experiment as the recommendation for titanium alloy cutting. The tool was clamped and rotated by and electric spindle Nakanishi HES 810-ST32 which the spindle speed is controlled and measured by Nakanishi E3000 spindle controller. The overhang length of the cutting tool is 10 mm to the spindle nose to prevent deformation and vibration from the cutting tool due to high cutting forces. Before the experiment, the workpiece and the cutting tool were first cleaned by using an ultrasonic cleaner for removing excessive dust that can cause build-up edges on the cutting tool. The cutting tool was then observed using a digital microscopes as shown in Figure 3 to verify the condition. A facing processes were also conducted to get common flat surface on the workpiece relative to the machine using a 3mm flat end mill.

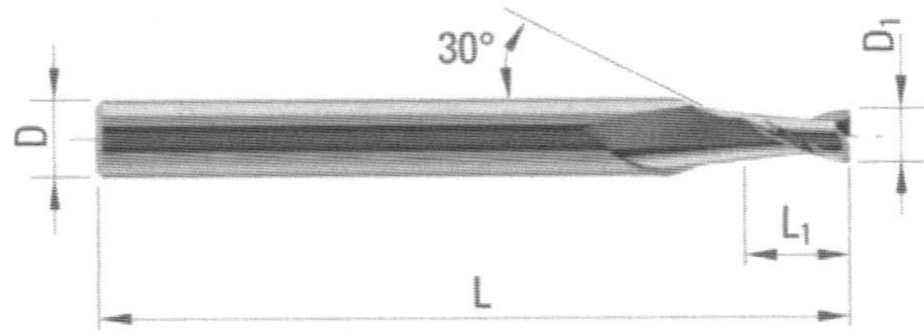

Figure 2. DIXI 7242 end mill cutting tool.

Table 1. DIXI 7242 specification.

Type	D1	D	L	L1	Number of flute	Material
End Mill	1 mm	3 mm	38 mm	2 mm	2	*Carbide*

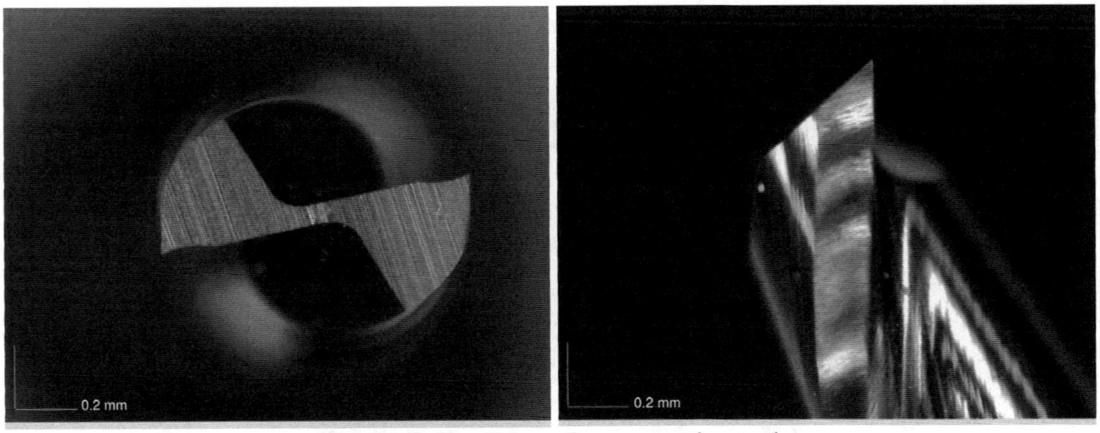

Figure 3. Condition of a new cutting tool.

In the experiments, slots with 4 mm length and 10 μm depth were produced by a slot-milling operation with the width equal to the cutting tool's diameter which is shown in Figure 4 . Zig toolpath was used in this experiment. The depth of cut for the slot-milling was constant at 10 μm. 9 different low speed cutting parameters were conducted for each slot as shown in Table 2. A new cutting tool is used for every slot with different cutting parameter to ensure the same cutting condition. Dry cutting was performed in this experiment. The output of the experiment is the surface roughness of each slot and all were measured using Surfcom 2900SD3. The measurement settings were 0.8 mm cut off, 2.4 mm evaluation length, and measure speed of 0.3 mm/s.

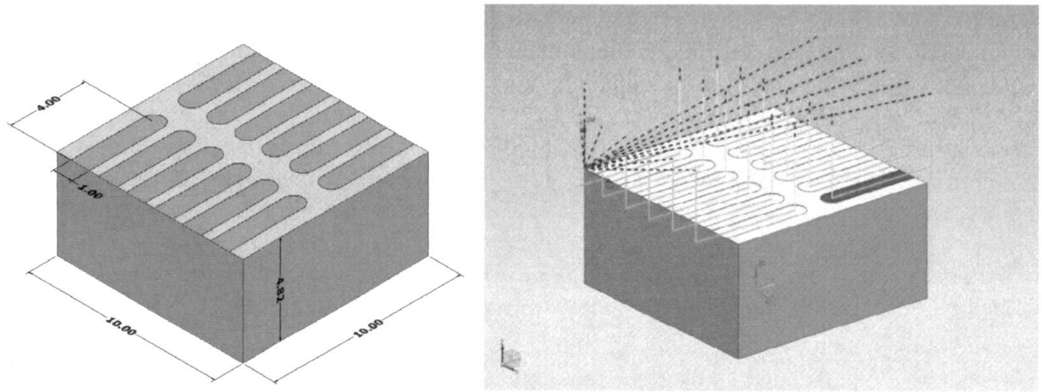

(a) **(b)**

Figure 4. (a) Design of slot milling (b) Slot milling toolpath.

Table 2. Variable of low speed cutting parameter.

Spindle Speed (rpm)	Feed Rate (mm/s)	Depth of Cut (μm)
	0.4	
3,000	0.8	
	1.6	
	0.4	
7,000	0.8	10
	1.6	
	0.4	
10,000	0.8	
	1.6	

3. Results and Discussion

The result of measured surface roughness from the various cutting parameters are shown in Table 3. The lowest surface roughness at 0.0972 μm was obtained by the lowest feed rate with the highest spindle speed. On the other hand, the highest surface roughness at 0.4496 μm was obtained by the highest feed rate with the lowest spindle speed. In order to give clearer visualization on the influence of cutting parameters on the machined surface roughness, the results of the experiment are shown in Figure 5 and Figure 6.

Table 3. Surface roughness of each cutting parameters.

Parameter	Spindle Speed (rpm)	Feed Rate (mm/s)	Surface Roughness (μm)
1		0.4	0.1522
2	3,000	0.8	0.2594
3		1.6	0.4496
4		0.4	0.1155
5	7,000	0.8	0.1546
6		1.6	0.1647
7		0.4	0.0972
8	10,000	0.8	0.1058
9		1.6	0.1446

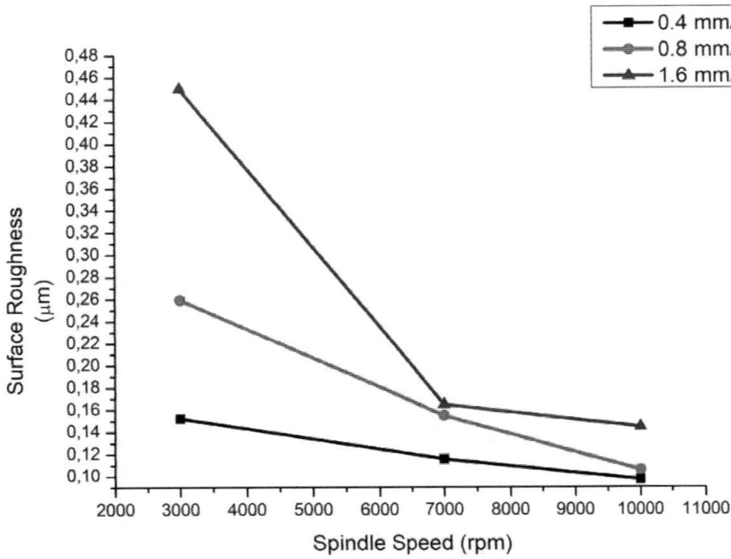

Figure 5. Influence of spindle speed to the surface roughness.

In Figure 5, it is seen that the surface roughness decreases as the spindle speed increases. The relationship shows that the increase of spindle speed causes the cutting forces to be decreased. This will lead to good tool life because of the low cutting load in the tool. Using a low feed rate with a higher spindle speed tend to produce lower surface roughness. Thus, using higher spindle speed with lower feed rate is recommended for saving cutting tools as well as producing a high surface quality.

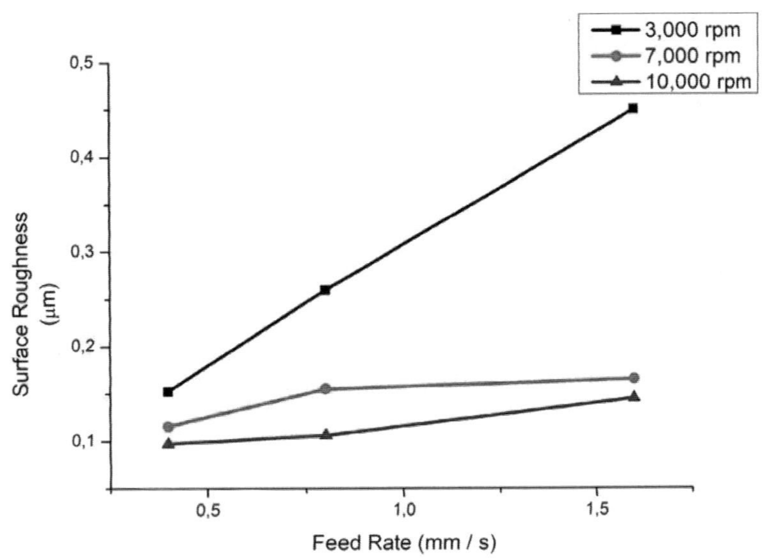

Figure 6. Influence of feed rate to the surface roughness.

In Figure 6, it is clearly seen that with the same spindle speed, the lower the feed rate, the lower the surface roughness is. This relationship shows that machining process in low feed rates allows the cutting tool to remove more material than the high feed rate does. High feed rate must be followed by higher spindle speed so that the removal of the material can be done more perfectly. If high feed rates are not followed by setting higher spindle speeds, the cutting load of the tool will be too much that can lead to excessive heat, rapid tool wear or even tool breakage that can crucially affects the surface roughness. It is clearly seen that the surface roughness is increasing significantly on the cutting parameter of the highest feed rate with the lowest spindle speed. Thus, the increase of feed rates must be followed by increasing spindle speed in order to lower the cutting load.

The difference of the surface view between the lowest and the highest surface roughness are shown in Figure 7 that are taken with digital microscope with 100× magnification. It is seen that the cutter marks on the lowest surface roughness of 0.0972 μm are thinner than on the highest surface roughness. It is shown that on the higher feed rate with low spindle speed tend to leave a thick mark on the surface that cause the surface roughness to rise because of the uncut material.

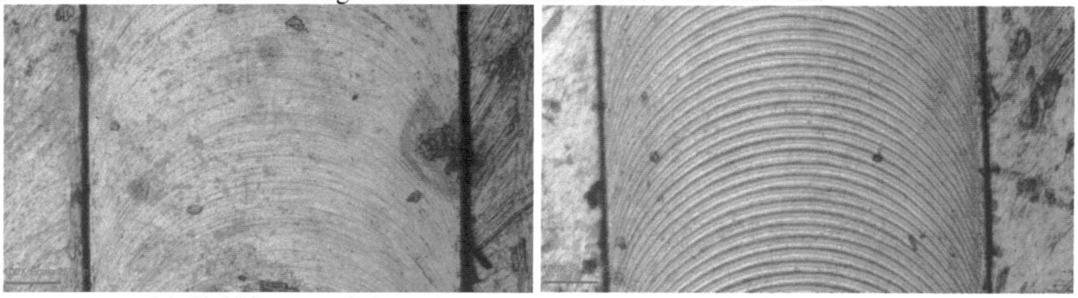

(a) 10,000 rpm – 0.4 mm/s **(b) 3,000 rpm – 1.6 mm/s**

Figure 7. Top view of the slot with (a) lowest and (b) highest surface roughness.

From the result of published researches of Ti-6Al-4V high-speed micro-milling, the surface roughness between the low-speed and high-speed Ti-6Al-4V does not making significant difference. It means that low speed micro-milling can be done in terms of saving cost of cutting tools and lower budget of machines without sacrificing the surface quality. In practice, despite of designing a production plan with a lowest machining time as possible, the cost and budgets for a production is still being considered. Thus, low speed machining Ti-6Al-4V micro-milling can still be a good choice for producing micro products with high surface quality in terms of low cost and budgeting.

4. Conclusion

From the results to the discussion of this experiment, it can be concluded that :

- Higher spindle speed in low-speed Ti-6Al-4V micro-milling can produces lower surface roughness.
- High feed rates in low-speed machining of Ti-6Al-4V micro-milling must be followed by higher spindle speed in order to lower the cutting load that causes tool wear or even tool breakage that can increases the surface roughness significantly.
- Cutting parameter with high feed rates with low spindle speed tend to leave thicker cutter marks that will increase the surface roughness
- The result between low-speed machining and high-speed machining of Ti-6Al-4V is not have significant differences
- Low speed machining Ti-6Al-4V micro-milling can still be a good choice for producing micro products with high surface quality in terms of low cost and budgeting.

Acknowledgments

This research was funded by PIT 9 Research Grant 2019-Universitas Indonesia with grant number NKB-0057/UN2.R3.1/HKP.05.00/2019.

5. References

[1]D. Huo, K. Cheng, and F. Wardle, 2010, *Design of a five-axis ultra-precision micro-milling machine-UltraMill. Part 1: Holistic design approach, design considerations and specifications*, *Int. J. Adv. Manuf. Technol.*, vol. 47, no. 9–12, pp. 867–877.

[2]M. Niinomi, 2003, *Recent research and development in titanium alloys for biomedical applications and healthcare goods*, *Sci. Technol. Adv. Mater.*, vol. 4, no. 5, pp. 445–454.

[3]T. Thepsonthi and T. Özel, 2014, *An integrated toolpath and process parameter optimization for high-performance micro-milling process of Ti–6Al–4V titanium alloy*, *Int. J. Adv. Manuf.*

Technol., vol. 75, no. 1–4, pp. 57–75.

[4]D. Biermann and A. Baschin, 2009, *Influence of cutting edge geometry and cutting edge radius on the stability of micromilling processes, Prod. Eng.*, vol. 3, no. 4–5, pp. 375–380.

[5]C. Bandapalli, B. M. Sutaria, D. V. Prasad Bhatt, and K. K. Singh, 2018, *Tool Wear Analysis of Micro End Mills - Uncoated and PVD Coated TiAlN; AlTiN in High Speed Micro Milling of Titanium Alloy - Ti-0.3Mo-0.8Ni, Procedia CIRP*, vol. 77, no. Hpc, pp. 626–629.

[6]W. Wang, S. H. Kweon, and S. H. Yang, 2005, *A study on roughness of the micro-end-milled surface produced by a miniatured machine tool, J. Mater. Process. Technol.*, vol. 162, pp. 702–708.

[7]A. Dadgari, D. Huo, and D. Swailes, 2018, *Investigation on tool wear and tool life prediction in micro-milling of Ti-6Al-4V, Nanotechnol. Precis. Eng.*, vol. 1, no. 4, pp. 218–225.

[8]E. Kuram and B. Ozcelik, 2015, *Optimization of machining parameters during micro-milling of Ti6Al4V titanium alloy and Inconel 718 materials using Taguchi method, Proc. Inst. Mech. Eng. Part B J. Eng. Manuf.*, vol. 231.

[9]P. Bach, G. Trmal, P. Zeman, J. Vana, and J. Maly, 2012, *High performance titanium milling at low cutting speed, Procedia CIRP*, vol. 1, no. 1, pp. 226–231.

[10]G. Kiswanto, D. L. Zariatin, and T. J. Ko, 2015, *The effect of spindle speed, feed-rate and machining time to the surface roughness and burr formation of Aluminum Alloy 1100 in micro-milling operation, J. Manuf. Process.*, vol. 16, no. 4, pp. 435–450.

IOP Conf. Series: Materials Science and Engineering **654** (2019) 012014 doi:10.1088/1757-899X/654/1/012014

The Effect of Machining Parameters to the Surface Roughness in Low Speed Machining Micro-milling Inconel 718

Gandjar Kiswanto[1], M Azmi[1], A Mandala[1] and T. J. Ko[2]

[1]Departement of Mechanical Engineering, Universitas Indonesia, Indonesia
[2]School of Mechanical Engineering, Yeungnam University, Republic of Korea

gandjar_kiswanto@eng.ui.ac.id

Abstract. In recent years, the rapid technology development in industry sector, especially in aviation industry has been increasing. So that we need more capabilities in the micro-scale manufacturing process, especially micro-milling process which can produce good surface roughness and high complexity parts. In this study, the effect of machining parameters to the surface roughness investigated by using Inconel 718 material with cutting tool diameter 1 mm, carbide material with coating TiAlN. The machining process was performed in low speed machining category. There are three variations of spindle speed (3.000, 7.000, 10.000 RPM) and also feed rate (0.5, 1, 2 mm/s) with constant depth of cut 10 μm. The relationship between machining parameters and surface roughness was obtained. The higher the feed rate, the higher the surface roughness produced. On the other hand, the higher the spindle speed the smaller the surface roughness. It also found that the machining process with low spindle speed, below 10.000 RPM, successfully carried out on hard-to-cut material, Inconel 718, that will be increasing the flexibility of machining parameters.

1. Introduction

Development of manufacturing process is necessary because of the rapid technology development, especially aviation industry. For example, the development of aircraft engine to achieve higher efficiency. Hard materials like Inconel 718 has been used as a material of aircraft engine parts including blades, disc, etc. Besides aircraft engines, Inconel 718 also used for rocket parts, gas turbine, and also molding because it has good strength, corrosion resistance, and also good creeping resistance at high temperatures. Inconel 718 consist of nickel, chromium, iron and etc that makes the material strong. However Inconel 718 has poor machinability because of its difficult-to-cut materials, so it needs more investigation for this material especially in manufacturing process. Optimization of electrochemical micro-machining has been done on Inconel 718 [1]. Drilling on Inconel 718 by using electrochemical machining has also been done. But, there is still difficulty on micro-machining Inconel 718 [2].

Aviation part's, like aircraft engine, gas turbine's blade, and also molding have a high complexity in shape to achieve optimal performance of the part's and also require high accuracy geometry. To fulfil these requirements, micro-milling process employed. Micro-milling is a part of micro-manufacturing process that has an ability of producing 2D and 3D parts with high complexity shape. It's also offers good surface quality and high accuracy after machining so it can deny finishing stage process like grinding and polishing. Removing this finishing stage process requires machining that results good surface quality, like surface roughness, and good geometry accuracy. So that to achieve these requirements, optimal machining process must be observed. Not only the rigidity of the machine

Content from this work may be used under the terms of the Creative Commons Attribution 3.0 licence. Any further distribution of this work must maintain attribution to the author(s) and the title of the work, journal citation and DOI.
Published under licence by IOP Publishing Ltd

tool, choosing optimal machining parameters significantly impact the surface quality, especially the surface roughness. Machining parameters consist of cutting tool, feed rate, spindle speed, depth of cut, etc. The effect of tool wear through machining process also must be considered since it can affect the cutting process when the cutting tool contact with workpiece material. Tool wear rate also will affect the cost of production of the product. Kiswanto et al (2015) investigated the effect of cutting parameters, like spindle speed, feed rate, and machining time, to the surface roughness. It was found that feed rate and machining time gave an impact significantly to the surface roughness. The lower feed rate would produce smoother surface [3].

The effect of machining parameters to surface quality in micro-milling Inconel 718 has been investigated in the past years. Kuram et al (2019) investigated the effect of overhang length of cutting tool and also feed per tooth to the cutting force during machining process. It was found that, overhang length affect cutting force and also produced chatter vibration. The longer of overhang length the higher of cutting force it be. The increasing of feed per tooth also increased cutting force because the enlargement of chip cross sectional area to be cut [4]. Sredanovic et al (2017) also investigated the machinability of Inconel 718. It was found that, Inconel 718 can be machined by using long neck flat micro end mills with diameter 0,6 mm. The best machining parameters combination obtained with feed per tooth 0.012 mm, cutting speed 40m/min, and depth of cut 0,02 mm. These parameter combination gave best output, especially in surface roughness. It was also found that the increasing of feed per tooth would increase surface roughness [5]. Kuram et al (2015) also investigated the effect of spindle speed, feed rate, and depth of cut. It was found that spindle speed and feed rate were a significant factor to surface roughness and also tool wear [6]. Lu et al (2018) optimized machining parameters with high material removal rate (MRR) to achieve machining efficiency [7].

Unfortunately, the recent research only covered spindle speed above 10.000 RPM and not yet covered below 10.000 RPM. In machining process, machining can be done with low spindle speed range because of the low budget specification of the machine tool, where the spindle speed only offers up to 10.000 RPM for machining process. Therefore, this research investigates the ability of machining Inconel 718 below 10.000 RPM and also the effect of machining parameters to the surface roughness. The relationship of machining parameters to the surface roughness with three variation of spindle speed and feed rate, and also constant depth of cut will be observed.

2. Experimental Setup

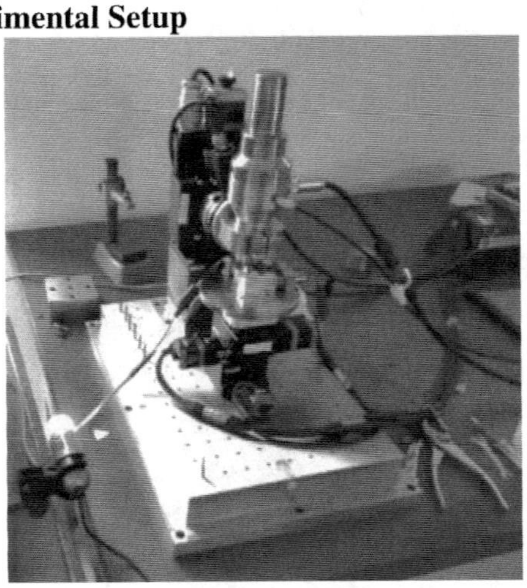

Figure 1. A miniaturized 5-axis micro-milling machines.

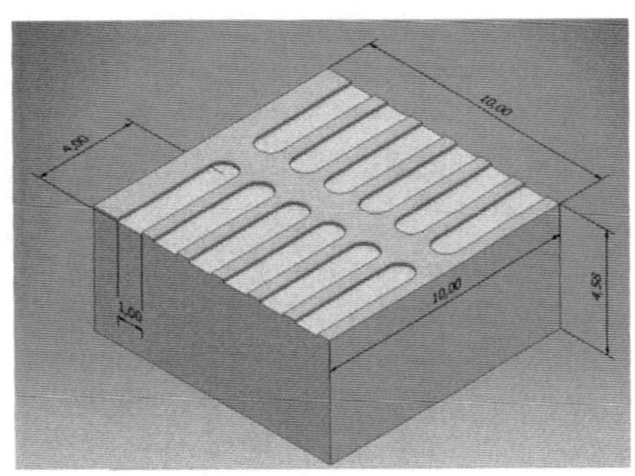

Figure 2. CAD for experimental of this research.

A miniaturized 5-axis micro-milling machine was used in this experiment for machining process. The machine tool movement was moved by motor stepper that has resolution 1 micron. The 3-Axis is linear axis that do linear movement and the 2-axis is rotational axis that do rotational movement. These motor stepper controlled with three units DS102 Suruga Seiki. An electric motor spindle

Nakanishi HES 810 also used in this experiments as a motor that provide spindle speed with maximum rotation 80.000 RPM. The picture of machine tool can be seen in Figure 1.

A carbide cutting tool DIXI 7242 with diameter 1 mm and 2 flutes with TiAlN coating was used in this experiments. Then, Inconel 718 with dimension 10 mm x 10 mm x 4,8 mm with hardness 66 HRC was selected as a workpiece materials. Before machining process, cleaning of workpiece material and cutting tool was done by using ultrasonic cleaners to remove dust and dirt. Then, facing process was employed in this experiment to get common flat surfaces as machining surface reference on the workpiece experiments. Each of machining parameters was machined in every channels which can be seen in Figure 2. Every machining parameters has been done by using new cutting tool to diminish tool wear effect in every machining parameters. The new tool condition can be seen in Figure 3.

In this experiments, the variation of machining parameters can be seen in Table 1. The low spindle speed selection below 10.000 RPM to achieve the aim of this research. Dry cutting without cutting fluid performed in this machining process. Then, overhang length of cutting tool was 15 mm to prevent high cutting force and chatter vibration during machining process. This experiments performed slot milling process. The surface roughness of every channels measured by using SURFCOM 2900 SD3 with cut off 0,8 mm, evaluation length 2,4 mm, and measure speed 0,3 mm/s.

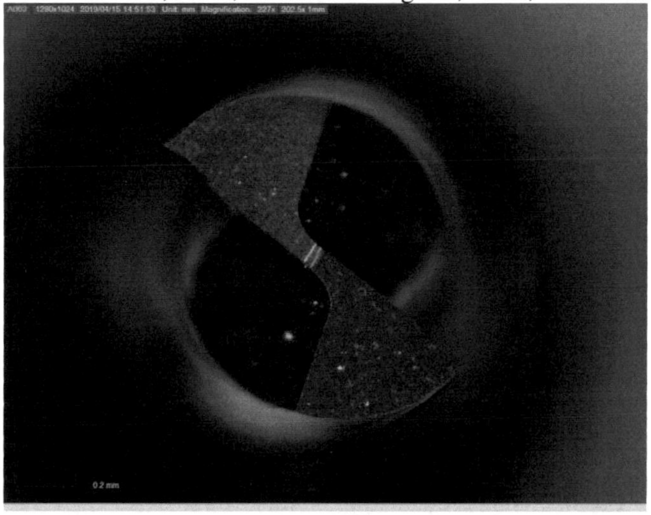

(a)

(b)

Figure 3. A condition of new cutting tool DIXI 7242 captured with dino-lite.

IOP Conf. Series: Materials Science and Engineering **654** (2019) 012014 doi:10.1088/1757-899X/654/1/012014

Table 1. Machining parameters of this experiments.

Parameter	Spindle Speed (RPM)	Cutting Speed (m/min)	Feed Rate (mm/s)	Feed per tooth (mm/tooth)
1	3000	9,43	0.5	0,0050
2	7000	22,00	0.5	0,0021
3	10000	31,43	0.5	0,0015
4	3000	9,43	1	0,0100
5	7000	22,00	1	0,0043
6	10000	31,43	1	0,0030
7	3000	9,43	2	0,0200
8	7000	22,00	2	0,0086
9	10000	31,43	2	0,0060

3. Results and Discussion

Every channel that has been machined measured by using SURFCOM 2900SD3 to get the surface roughness (R_a).The result of surface roughness every machining parameters can be seen in Table 2.

Surface quality, like surface roughness and burr, influenced by machining parameters. Based on Figure 4, we can see the effect of feed rate on surface roughness. If the value of feed rate was increasing, the surface roughness in the channel was also increasing. Machining parameters with spindle speed 3.000 RPM and feed rate 2 mm/s produced highest surface roughness with value 0.4163 μm. Otherwise, machining parameters with spindle speed 10.000 RPM and feed rate 0.5 mm/s produced lowest surface roughness with value 0.0673 μm. So that low spindle speed with high feed rate will produce high surface roughness and vice versa.

Table 2. Results of surface roughness of each machining parameters.

Parameter	**Spindle Speed (RPM)**	**Cutting Speed (m/min)**	**Feed Rate (mm/s)**	**Feed per tooth (mm/tooth)**	**Surface Roughness (μm)**
1	3000	9,43	0.5	0,0050	0,1253
2	7000	22,00	0.5	0,0021	0,0678
3	10000	31,43	0.5	0,0015	0,0673
4	3000	9,43	1	0,0100	0,2974
5	7000	22,00	1	0,0043	0,0874
6	10000	31,43	1	0,0030	0,0935
7	3000	9,43	2	0,0200	0,4163
8	7000	22,00	2	0,0086	0,0932
9	10000	31,43	2	0,0060	0,1011

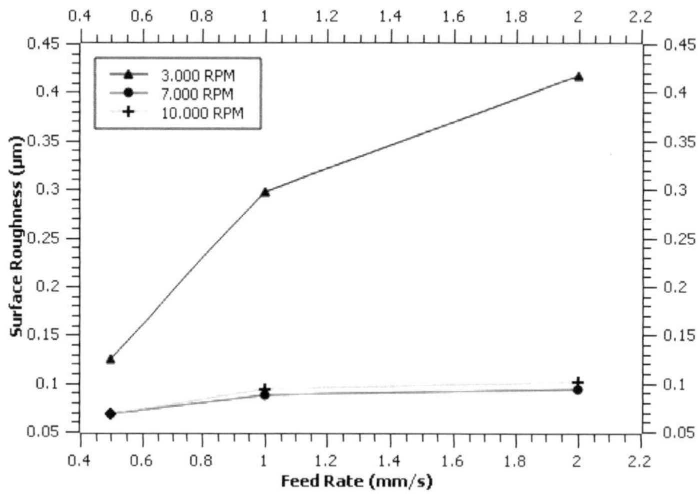

Figure 4. The influences of feed rate into surface roughness.

Feed rate refers to the velocity of workpiece relative to the cutting tool. The higher value of feed rate, the faster machining process will be. Short machining process will cost into tool wear that can affect surface quality. Feed rate has a linear relationship with feed per tooth. The value of feed per tooth in this experiments can be seen in Table 1. Feed per tooth is the magnitude of material that will be cut by every tooth of the cutting tool. If feed per tooth was increasing, the cutting load in the machining process was also increasing because of the enlargement of the cross sectional area of chip. This will affect the tool wear that will disturb cutting process and then gave bad impact to the surface roughness. The increasing of feed per tooth will lead into the escalation of tool wear rate. At spindle speed 3.000 RPM, we can see clearly that the increasing of feed rate from 0.5 mm/s to 2 mm/s produced high surface roughness. After that, the effect of increasing of spindle speed on surface roughness can be seen in Figure 5.

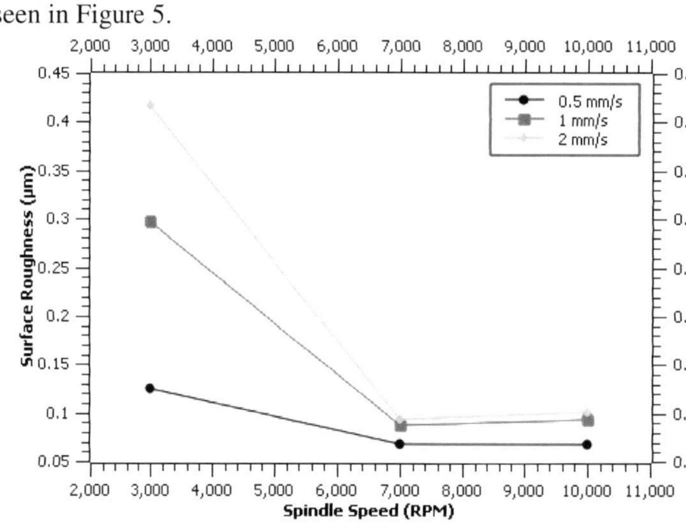

Figure 5. The influences of spindle speed into surface roughness.

Based on Figure 5, if the rotation of spindle speed is increasing, the surface roughness is decreasing. This occurred at the same feed rate with the increasing of spindle speed rotation. It can be seen in Table 1, the feed per tooth will be decreasing with the increasing of spindle speed. So the cutting load of the every tooth in the cutting tool will be decreasing. Spindle speed also refers into cutting speed. The increasing of cutting speed will reduce cutting force. But, high cutting speed will lead to high friction resulting in an increase in temperature of the cutting tool. This will lead to the increasing of tool wear that will affect surface roughness. However, this would have significant impact if we done high cutting speed machining with longer machining time.

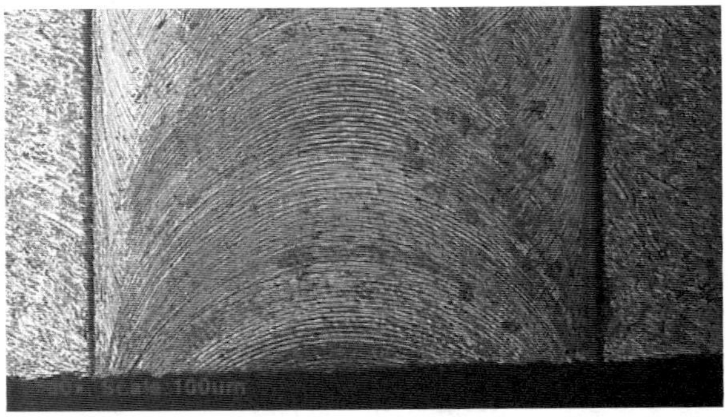

Figure 6. The surface condition that has been machined with parameter 1.

The appearances of built up edge (BUE) during machining process was found in this experiment. Based on Figure 6, parameter 1, there was a chip attached to the surface after the machining process was carried out. This chip on the surface of the channel will affect the surface roughness measurement. The stylus will detect the chip as a surface and then increase the surface roughness of the channel. The appearances of built up edge must be avoided since it can affect the surface quality and cutting process.

4. Conclusion

This research observed and evaluated the effect of machining parameters into surface roughness. So this can be concluded that:

- Low surface roughness achieved by using machining parameters combination with spindle speed 10.000 RPM and feed rate 0,5 mm/s.
- High feed rate will produce higher surface roughness. It occurs because of the increasing of cutting load in machining process.
- High spindle speed will produce lower surface roughness. It occurs because of the increasing of spindle speed that lead into the decreasing of cutting load where the cutting force will be decreasing also.
- Machining process with low spindle speed, below 10.000 RPM, successfully carried out on hard-to-cut material, Inconel 718. This will increase the flexibility of the selection of machining parameters, especially spindle speed. Where the surface roughness can be produced below 0.5 μm.

Acknowledgements

This research is funded by PIT 9 Research Grant 2019 – Universitas Indonesia with number NKB-0057/UN2.R3.1/HKP.05.00/2019.

5. References

[1] T Geethapriyan, K Kalaichelvan, & T Muthuramalingam.(2016).Multi Performance Optimization of Electrochemical Micro-Machining Process Surface Related Parameters on Machining Inconel 718 using Taguchi-grey relational analysis. *La Metallurgia Italiana*, 2016(4), 13-19.

[2] T Geethapriyan, K Kalaichelvan, & T Muthuramalingam .(2016).Influence of coated tool electrode on drilling Inconel alloy 718 in Electrochemical micro machining. Procedia CIRP, 46(1), 127-130.

[3] Kiswanto, G., Zariatin, D. L., & Ko, T. J. (2015). The effect of spindle speed, feed-rate and machining time to the surface roughness and burr formation of Aluminum Alloy 1100 in micro-milling operation. Journal of Manufacturing Processes, 16(4), 435–450. https://doi.org/10.1016/j.jmapro.2014.05.003

[4] Kuram, E. (2019). Overhang length effect during micro-milling of Inconel 718 superalloy. Journal of the Brazilian Society of Mechanical Sciences and Engineering, 41(4), 1–12. https://doi.org/10.1007/s40430-019-1669-z

[5] Kuram, E., & Ozcelik, B. (2015). Optimization of machining parameters during micro-milling of Ti6Al4V titanium alloy and Inconel 718 materials using Taguchi method. Proceedings of the

Institution of Mechanical Engineers, Part B: Journal of Engineering Manufacture, 231(2), 228–242. https://doi.org/10.1177/0954405415572662

[6] Sredanovic, B., Lakic, G., Kramar, D., & Kopac, J. (2017). Study on the Machinability Characteristics of Inconel 718 Super Alloy During Micro-Milling. Proceedings of 5th International Conference on Advanced Manufacturing Engineering and Technologies, 375-385. doi:10.1007/978-3-319-56430-2_28

[7] Lu, X., Zhang, H., Jia, Z., Feng, Y., & Liang, S. Y. (2018). Cutting parameters optimization for MRR under the constraints of surface roughness and cutter breakage in micro-milling process. Journal of Mechanical Science and Technology, 32(7), 3379–3388. https://doi.org/10.1007/s12206-018-0641-7

IOP Conf. Series: Materials Science and Engineering **654** (2019) 012015 doi:10.1088/1757-899X/654/1/012015

Development of Langevin Piezoelectric Transducer-based Two Dimensional Ultrasonic Vibration Assisted Machining (2D UVAM) on 5-axis Micro-milling Machine

Gandjar Kiswanto[1], Poly[1], Y R Johan[1], T. J. Ko[2], R Kurniawan[2]

[1]Department of Mechanical Engineering, Universitas Indonesia, Kampus Baru UI, Depok 16424, Indonesia
[2]School of Mechanical Engineering, Yeungnam University, 280 Daehak-ro, Gyeongsan-si, Gyeongsangbuk-do 38541, Republic of Korea

gandjar_kiswanto@eng.ui.ac.id

Abstract. The purpose of this study is to design and develop a Two-Dimensional Ultrasonic Vibration Assisted Micro-milling (2D UVAMM) system. Two-dimensional ultrasonic vibrations produced by 2D UVAMM are used to vibrate the workpiece in the micro-milling process. The system is developed using the principle of two Langevin piezoelectric transducers which have the ability to produce ultrasonic vibrations with small vibration amplitude. Process of 2D UVAMM design optimization was done by modal simulation using Finite Element Analysis method. Both Langevin piezoelectric transducers are designed to have symmetrical and asymmetrical vibration modes with the same natural frequency, so that elliptical pattern vibrations can be generated on the workpiece. The 2D UVAMM system operates at a natural frequency of 24 kHz and has an estimated total displacement on the normal and tangential axes respectively 0.766 µm and 0.382 µm. Two power sources with frequency of 24 kHz, phase difference of 90 degrees, and peak-peak voltage of 212 volt were supplied by an ultrasonic generator to excite both of the Langevin piezoelectric transducers. To confirm the developed 2D UVAMM system, experiments were conducted to compare the surface roughness of Aluminum 6061-T6 through micro-milling with conventional method and with additional of the 2D UVAMM system.

1. Introduction

In the current era, almost all tools used by humans are minimalist and portable devices, due to the miniaturization of components or parts. For example, the automotive industry, aerospace industry, medical field and energy industry have conditioned the need for parts with relatively small dimensions [1]. Increasing demand for components and micro-products requires the improvement and development of existing manufacturing technologies so as to enable faster and more reliable micro components production. To overcome this problem, now micro components can be easily produced through a machining process known as micro-machining.

Nowadays there are many micro-machining techniques that have been applied in industries, but among them the micro-milling technique is one that is widely used. Micro-milling is a machining technique for producing micro components using a sharp cutting tool that rotates to cut or scrape the material clamped on the workbench with material disposal in the form of chips [2]. Micro-milling enables industries to produce micro components with complex shapes and patterns.

Content from this work may be used under the terms of the Creative Commons Attribution 3.0 licence. Any further distribution of this work must maintain attribution to the author(s) and the title of the work, journal citation and DOI.
Published under licence by IOP Publishing Ltd

However the industries mentioned above have used a lot of hard and brittle materials, such as ceramics and glass, and conventional micro-machining processes are not enough to obtain machining results with high level quality of tolerance and surface roughness. Various efforts have been made to improve the quality of the micro machining process, until a method is found where the machining process is equipped with a vibration generating device [3]. In the machining process, giving vibrations can increase machining results when applied with the correct method and parameter [4]. Vibration assisted machining (VAM) is a cutting method in which vibrations with a certain frequency and amplitude are given to the cutting tool or to the workpiece, in addition to the relative motion between the cutting tool and workpiece, to obtain better cutting performance. The basic principle of vibration assisted machining (VAM) is the occurrence of contacts separation between the cutting tool and workpiece periodically [5].

In vibration assisted machining (VAM) process, a small amplitude and high frequency vibration source is used [6]. This can be achieved by using an actuator called piezoelectric actuator. Piezoelectric actuators are composed of a ceramic material that can contract and expand when an electrical voltage is applied. Because of its ability, the piezoelectric actuator is used to produce small displacements with high frequency and response speed. Compared to other types of actuators, piezoelectric actuators also generates high force or strength relative to their small size [7].

The micro-milling process can also be equipped with additional vibration device, so it is called vibration assisted micro-milling (VAMM). In VAMM process, vibration motion can be given linearly in one direction (1D VAMM) or elliptically in two directions (2D VAMM) [6]. Several 2D VAMM systems have been developed previously using linear piezoelectric actuators with frequency ranges that were still below ultrasonic [5,8]. Chen et al. [3] states that to date the 2D VAMM system in the micro-milling process is limited to frequencies below 3 kHz, which is not suitable for the micro-milling process. It was also stated that in order to achieve proper periodic chip separation in one workpiece engagement cycle, ultrasonic frequencies with range above 20 kHz are needed. Therefore, the development of the Two Dimensional Ultrasonic Vibration Assisted Micro-milling (2D UVAMM) system is a challenge nowadays.

Based on the problems mentioned above, the aim of this research is to develop a 2D UVAMM system in order to produce ultrasonic vibrations for micro-milling process using Langevin piezoelectric transducer which have the ability to produce vibrations with higher frequencies compared to linear piezoelectric actuator. From the results of the development, an experiment will be conducted to compare the surface roughness of machined products from conventional method micro-milling and micro-milling equipped with the developed 2D UVAMM system.

2. Langevin Piezoelectric Transducer

Ultrasonic wave applications have been developed over the past few decades and piezoelectric ceramic materials have been commonly used as the source of these waves [9]. One example of the application of piezoelectric material in this field is the Langevin piezoelectric transducer. This type of transducer has a construction consisting of one or more piezoelectric elements which are clamped between two pieces of metal. Both pieces of metal are often known as mass or end mass.

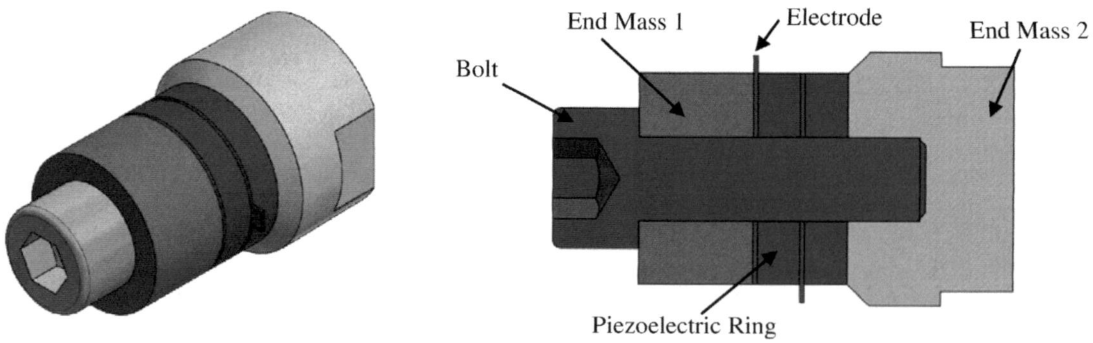

Figure 1. Construction and cross section of Langevin piezoelectric transducer

The illustration shown in Figure 1 above is a general construction and the cross section of the Langevin piezoelectric transducer. Both piezoelectric elements are compressed between the two end masses which are separated by electrodes using a bolt through the middle part of the two masses. End mass 2 is usually made of lightweight metal and end mass 1 is made of heavy metal to ensure that the energy produced by the piezoelectric elements can be maximally transferred to the cross section associated with end mass 2 [10].

After being connected to a power source with a certain frequency, the transducer will then resonate. A Langevin piezoelectric transducer will resonate at a certain resonance frequency and amplitude. By changing the dimensions and geometries of the end mass on the transducer, the transducer can be adjusted for applications with the desired frequency and amplitude [10-11]. Abdullah et al. [9] stated that designing the Langevin piezoelectric transducer was not easy, because there was no valid information and equation to design and analyse the dynamic behaviour of the ultrasonic transducer. Therefore the Finite Element Analysis (FEA) method is one of the reliable analysis methods for analysing ultrasonic transducer.

3. Design and Optimization of 2D UVAMM System

3.1. Design Limitation

In designing the 2D UVAMM system, attention needs to be paid to the working envelope dimensions of the micro-milling machine. The dimensions of the 2D UVAMM system must be proportional so as not to disturb the movement of the machine. The micro-milling machine used in this study is Hadia Micromill-5X, a miniaturized table/spindle tilting type micro-milling machine shown in Figure 2. It consists of five axes of movement, with three XYZ linear axes and two AC rotation axes. Each axis is driven by a stepper motor stage which is controlled by DS102 Suruga Seiki stepper motor driver. A Nakanishi HES 810-ST32 electric motor spindle is used to rotate the cutting tool to a maximum speed of 80,000 RPM. The 2D UVAMM system developed in this study was then used to produce vibrations that oscillate on two axes that are level with the planar surface of the workpiece, it will then be placed on the C axis motor stage as shown in Figure 3.

Figure 2. Hadia Micromill-5X

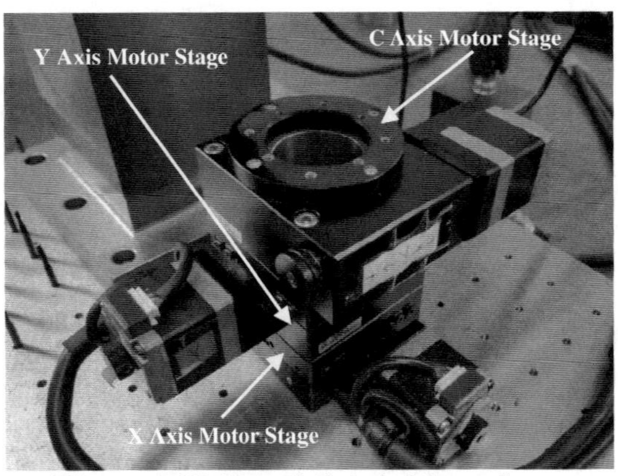

Figure 3. X, Y, and C Axis Motor Stage

3.2. Design of 2D UVAMM

The design of the 2D UVAMM system was developed using the Langevin piezoelectric transducer principle. This design was made based on a reference from Kurniawan et al. [12], a surface texturing device using two-dimensional ultrasonic vibrations. This system consists of two bolt-clamped Langevin piezoelectric transducers which are installed with a 90 degree angle difference. The reason for choosing this type of transducer is because of its ability to produce ultrasonic vibrations with small amplitude that is still within the micrometer range. Each Langevin piezoelectric transducer used is composed of four piezoelectric ring actuators. The piezoelectric ring used in study is produced by Soar Piezo, with an outer diameter of 15 mm, inner diameter of 6 mm, thickness of 3 mm, and it is made of PZT4 material. Detailed specifications of the piezoelectric ring used are shown in Table 1.

Table 1. Piezoelectric Ring Specification

E	C PF	T_c °C	Tg δ %	Fr kHz	Kp %	Zr Ω	Q_m	d_{33} m/V
1500	520	300	0.5	107	43	30	600	320×10^{-12}

Previously, the design of Kurniawan's Langevin piezoelectric transducer [12] was used to vibrate the head block installed with a Polycrystalline Diamond (PCD) cutting tool. Whereas in the purpose of this study, a 2D UVAMM system was developed to generate vibrations for the workpiece in micro-milling process, therefore the head block needs to be modified into a workpiece clamp which can simultaneously receive vibrations produced by both Langevin piezoelectric transducers.

Three-dimensional modeling software used in designing the 2D UVAMM system is Autodesk Inventor 2016. CAD design of the 2D UVAMM can be seen in Figure 4. The design consists of two Langevin piezoelectric transducers which are mounted perpendicular to each other on a holder block made of AISI 1045 medium carbon steel. Both end mass with diameter of 15 mm which function as bolts to tighten the system are made of the same material. Head block functions as a workpiece clamp and it is made of Aluminum Duralumin 2024 Alloy which is lighter than steel, thus the resulting vibrations produced can be channeled effectively to the workpiece. The workpiece material used in this study is Aluminum 6061-T6. The whole system is then connected to the C axis motor stage using an AISI 1045 medium carbon steel base plate, and an Aluminum Duralumin 7075 Alloy base block.

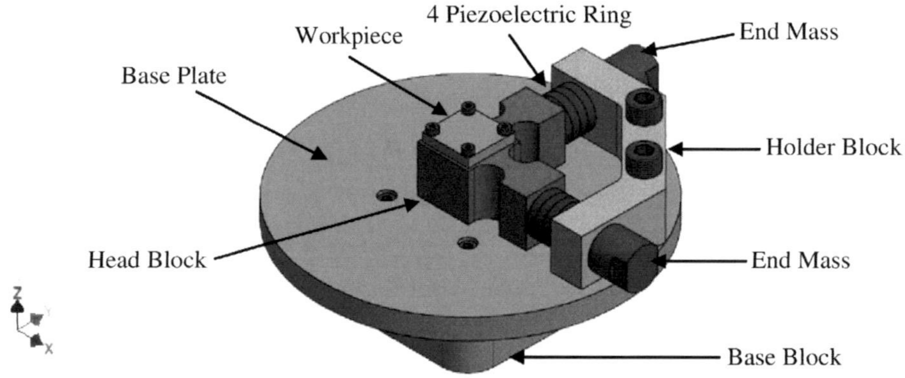

Figure 4. CAD Design of 2D UVAMM

3.3. Modal Simulation of 2D UVAMM

CAD design that were previously made need to be evaluated to analyze its dynamic behavior. The evaluation process is done by modal simulation through the Finite Element Analysis (FEA) method. ANSYS Workbench 17.0 software is used to determine the natural frequency of each vibration mode formed from the 2D UVAMM. Two vibration modes that will be used are the symmetrical and asymmetrical modes. The following Figure 5 shows the illustration of the two vibration modes.

When the two Langevin piezoelectric transducers are in the same phase, the system will produce a symmetrical vibration mode, whereas when both of the Langevin piezoelectric transducers are not in phase, an asymmetrical vibration mode will occur. Vibration motion to the normal axis from the tip of the transducer is produced by symmetrical mode, while the vibration motion to the tangential axis is caused by asymmetrical mode. The process of optimizing the design of 2D UVAMM system is done so that vibrations can be generated with an elliptical motion, where the phenomenon can be achieved if both symmetrical and asymmetrical vibration modes have a natural frequency that is almost close or equal. Kurosawa et al. [11] stated that the natural frequency of symmetrical and asymmetrical vibration modes can be modified by changing the dimensions and geometries of the end mass. Therefore, several simulation processes with different end mass length variations need to be carried out to identify the appropriate natural frequency in order that both vibration modes can be formed simultaneously on 2D UVAMM system.

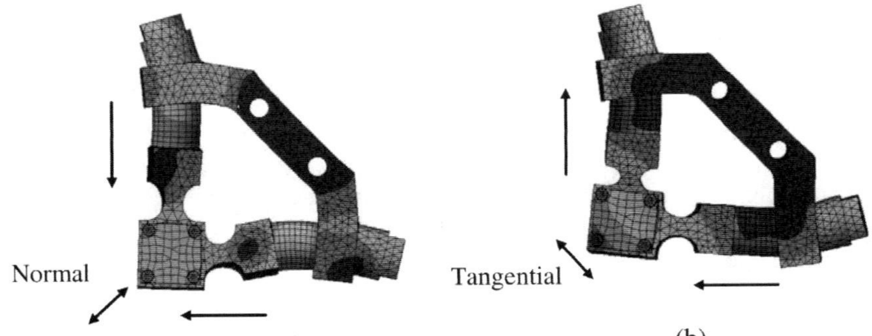

Figure 5. 2D UVAMM Vibration Modes (a) Symmetrical (b) Asymmetrical

4. Experiments and Results

4.1. Modal Simulation Results

End mass with a diameter of 15 mm with variations in length of 12 mm, 13 mm, 13.5 mm, 14 mm, and 15 mm are used in the modal simulation process to determine its effect on the natural frequency of both symmetrical and asymmetrical vibration modes. From the modal simulation through ANSYS Workbench 17.0 software, three pairs of symmetrical and asymmetrical vibration modes with different natural frequency ranges are obtained. Symmetrical and asymmetrical vibration modes 1, 2 and 3 are sorted by their natural frequency, detailed information is shown in Table 2 and Figure 6.

Table 2. 2D UVAMM Modal Simulation Results

End Mass Length mm	**Natural Frequencies**					
	Symmetric Mode 1 (Hz)	Asymmetric Mode 1 (Hz)	Symmetric Mode 2 (Hz)	Asymmetric Mode 2 (Hz)	Symmetric Mode 3 (Hz)	Asymmetric Mode 3 (Hz)
12	6,573	4,052	15,577	19,106	25,470	25,103
13	6,503	4,008	15,354	18,620	24,423	24,285
13.5	6,468	3,986	15,231	18,397	**23,969**	**24,047**
14	6,432	3,963	15,090	18,138	23,570	23,836
15	6,359	3,917	14,755	17,549	22,938	23,585

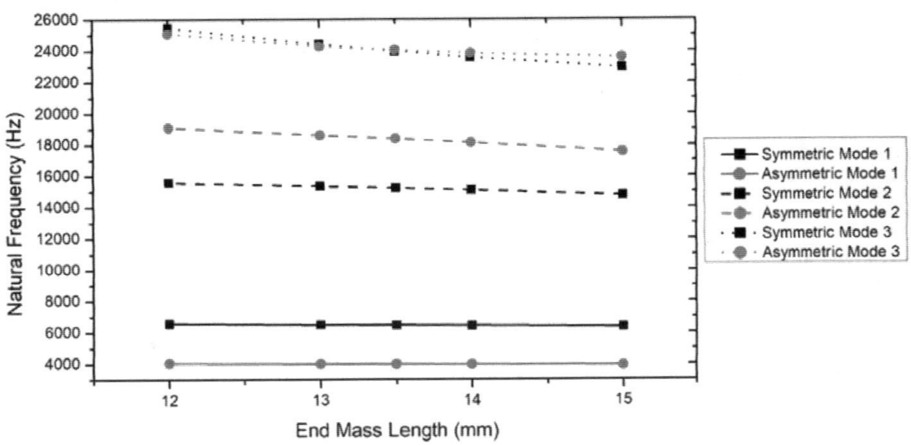

Figure 6. Effect of end mass length on 2D UVAMM vibration modes

It can be seen in Table 2 with the end mass length ranging from 12 mm to 15 mm, the natural frequencies of symmetrical vibration modes 1 occur at 6,359 Hz to 6,573 Hz. While the asymmetrical vibration modes 1 has natural frequencies ranging from 3,917 Hz to 4,052 Hz. The pattern obtained is that the natural frequency of each vibration mode will decrease as the end mass length increases. Symmetrical and asymmetrical vibration modes 1 cannot be used in research because the natural frequencies are still below the ultrasonic. Moreover, the natural frequencies differences between the two vibration modes are quite large, ranging from 2,442 Hz to 2,521 Hz.

With the same variation of end mass length, natural frequencies of symmetrical vibration modes 2 are obtained at 14,755 Hz to 15,577 Hz, while the natural frequencies of asymmetrical vibration modes 2 occur at 17,549 Hz up to 19,106 Hz. The same pattern is obtained with the vibration modes 1, an increase in end mass length will decrease the natural frequencies of both symmetrical and asymmetrical vibration modes. Same as before, the natural frequencies of the vibration modes 2 are still below ultrasonic, thus it cannot be used in this study. Natural frequencies differences between symmetrical and asymmetrical modes ranging from 2,794 Hz to 3,529 Hz are still large, so it can be concluded that elliptical vibration will not be formed.

Unlike the pairs of vibration modes 1 and 2, symmetrical and asymmetrical vibration modes 3 occur in the ultrasonic range. Symmetrical vibration modes 3 are obtained in the range of 22,938 Hz up to 25,470 Hz, while the natural frequencies of asymmetrical vibration modes 3 turn up in the range of 23,585 Hz to 25,103 Hz. Similar pattern is obtained, the addition of end mass length will also reduce the natural frequencies. In vibration modes 3, the differences between the natural frequencies of the symmetrical and asymmetrical modes are quite small, which ranges from 78 Hz to 647 Hz. The smallest difference in natural frequency occurs at the end mass length of 13.5 mm.

From the results of the modal simulation, it can be concluded that the pair of symmetrical and asymmetrical vibration modes 3 with end mass length of 13.5 mm can be used in this study. With that, natural frequency of the symmetrical vibration mode occurs at 23,969 Hz, and the asymmetrical mode at 24,047 Hz. Both natural frequencies are close to 24,000 Hz, hence the natural frequency value is used in the developed 2D UVAMM. Both vibration modes only have differences of 31 Hz and 47 Hz with the operating frequency of 24,000 Hz. Therefore the 2D UVAMM can produce symmetrical and asymmetrical vibration modes simultaneously, thus elliptical trajectory vibrations can be formed.

4.2. Preliminary Machining Results

The fabricated 2D UVAMM is then mounted on the C axis motor stage of Hadia Micromill-5X machine. Conventional machining process and machining with the addition of 2D UVAMM system were carried out to compare the surface roughness between the two. The experimental setup of the machining process is shown in Figure 7. The two Langevin piezoelectric transducer are excited by an ultrasonic generator from Hybrid Precision. Ultrasonic generator USG-110 was tuned to a frequency of 24 kHz according to the natural frequency of both symmetrical and asymmetrical vibration modes from the 2D UVAMM. The magnitude of phase difference used in both vibration modes is 90 degrees, whereas the output peak-peak voltage used is 212 Volt.

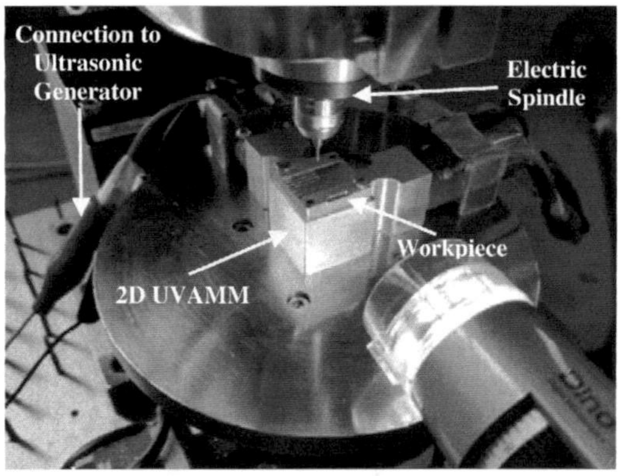

Figure 7. Machining experimental setup

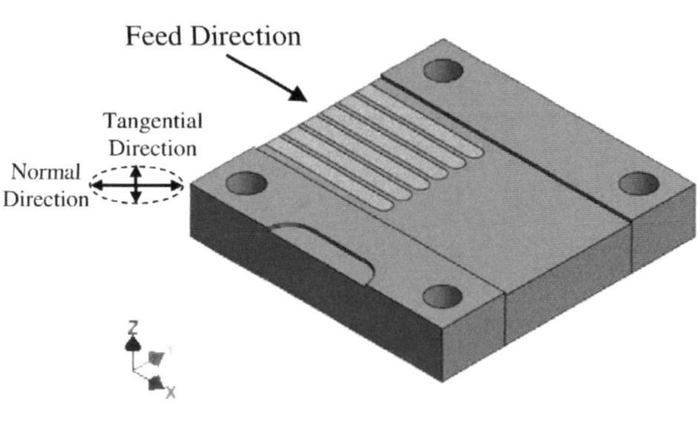

Figure 8. Feed and elliptical direction on workpiece

Surface roughness verification was done by slot cutting process using DIXI 7242 two flutes uncoated carbide endmill cutting tool, with diameter of 1 mm. Because the developed 2D UVAMM is still a initial prototype, Aluminum 6061-T6 is suitable to be selected as the test material in this study with the reason of its relatively soft material properties. In addition, Aluminum 6061-T6 is also widely used in industry today for the process of making lightweight aircraft and aerospace components. Slot cutting of 1 mm width, 10 μm depth, and 8 mm length channels were carried out on Aluminium 6061-T6 workpiece with dimension of 18 mm × 18 mm × 3.3 mm. Both conventional and 2D UVAMM equipped machining processes were treated with the same cutting parameters, including 7,000 RPM spindle rotation speed, feed rate of 0.2 mm/s, and 10 μm depth of cut. Each machining process is then repeated three times. For the machining process with 2D UVAMM, Figure 8 shows the feed direction towards the motion of oscillation from the resulting elliptical vibrations.

$$U = n.\ d_{33}.\ V_{P\text{-}P} \tag{1}$$

The value of the elliptical amplitude generated by the 2D UVAMM can be estimated using the amplitude generated by the both piezoelectric actuators on the X and Y axes according to Equation (1). Where n is the amount of stacked piezoelectric rings layers, d_{33} is the piezoelectric ring coefficient and $V_{P\text{-}P}$ is the supplied peak to peak voltage. Then by using the resultant and decomposition of vectors, it can be obtained the elliptical vibration amplitude against the normal and tangential axes. From the calculation results, the estimated total displacements produced by the 2D UVAMM on the normal and tangential axes are 0.766 μm and 0.382 μm, respectively.

Surfcom 2900SD3-12 measuring device was used to measure the surface roughness of the machined slots with a cut off value of 0.8 mm, measurement length of 4 mm, and measurement speed of 0.3 mm/s. From the measurement results, the average surface roughness in the conventional machining slots is equal to 0.1429 μm, whereas the average surface roughness in machined slots using 2D UVAMM drops to 0.1334 μm. The decrease in surface roughness is 6.626%. Ultrasonic vibrations produced by 2D UVAMM will create alternating oscillation movements on the workpiece, causing the cutting tool to re-cut the surface that has not been evenly cut before, thus improving surface quality. Shown in Figure 9, there are significant differences between the cutter marks of the two processes. Cutter mark produced through the 2D UVAMM machining process has a pattern like small scales.

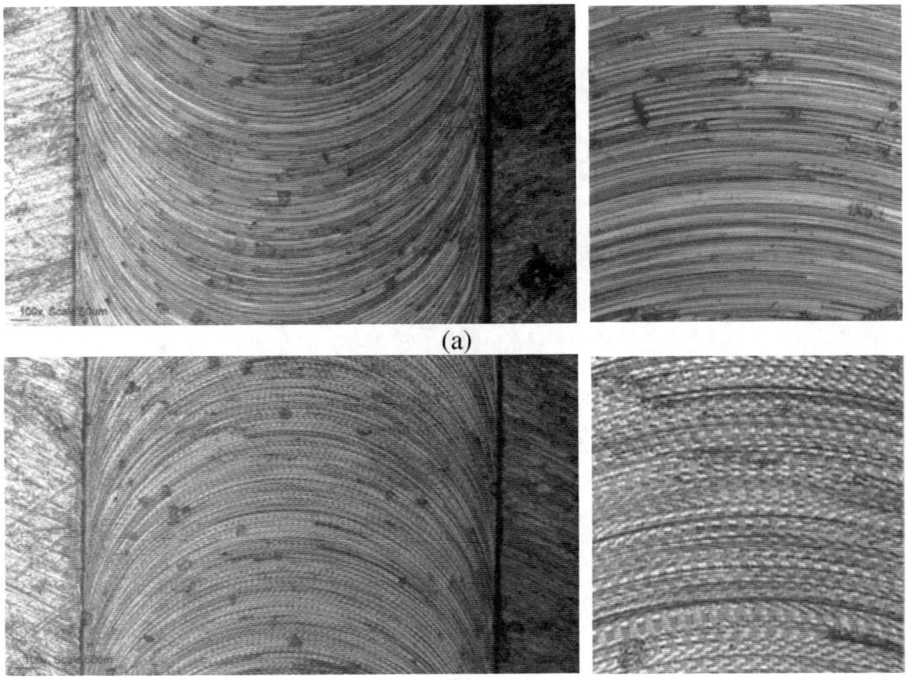

(a)

(b)

Figure 9. Surface cutter marks (a) Conventional micro-milling (b) Micro-milling with 2D UVAMM

5. Conclusions

After developing a Two Dimensional Ultrasonic Vibration Assisted Machining (2D UVAM) system on 5-axis micromilling machine and then used for Aluminum 6061-T6 machining process, the following conclusions can be drawn.

- The 2D UVAMM was developed using the principle of two Langevin piezoelectric transducers. The reason for choosing this type of transducer is because of its ability to produce ultrasonic vibrations with micro scale amplitude.
- Two symmetrical and asymmetrical vibration modes were used on the 2D UVAMM, vibration motion to the normal axis from the tip of the transducer is produced by symmetrical mode, while the vibration motion to the tangential axis is caused by the asymmetrical mode.
- Natural frequencies of both vibration modes were optimized by modifying the end mass geometry, an increase in end mass length will decrease the natural frequencies of both vibrations modes.
- Proposed 2D UVAMM system operates at an ultrasonic frequency of 24,000 Hz with estimated total displacement on the normal and tangential axes respectively 0.766 μm and 0.382 μm. Natural frequencies from both vibration modes are almost equal, thus an elliptical trajectory vibrations can be formed on the workpiece.
- Based on micro-milling machining experiments, the elliptical ultrasonic vibrations produced by 2D UVAMM on Aluminum 6061-T6 workpiece helps improve surface quality.

Acknowledgements

This research was funded by PIT 9 Research Grant 2019 - Universitas Indonesia with grant number NKB-0057/UN2.R3.1/HKP.05.00/2019.

6. References

[1] Sredanović B, Globočki-Lakić G, Kramar D and Pušavec F 2018 Influence of workpiece hardness on tool wear in profile micro-milling of hardened tool steel *Tribology in Industry* **40** 100-107
[2] Alting L, Kimura F, Hansen H N and Bissacco G 2003 Micro engineering *CIRP Annals* **52** 635-657
[3] Chen W, Huo D, Shi Y and Hale J M 2018 State-of-the-art review on vibration-assisted milling: principle, system design, and application *The International Journal of Advanced Manufacturing Technology* **97** 2033–49
[4] Ibrahim R 2010 Vibration assisted machining: modelling, simulation, optimization, control and applications *Brunel University School of Engineering and Design PhD Thesis*
[5] Chern G L and Chang Y C 2006 Using two-dimensional vibration cutting for micro-milling *International Journal of Machine Tools and Manufacture* **46** 659-666
[6] Brehl D E and Dow T A 2008 Review of vibration-assisted machining *Precision Engineering* **32** 153-172
[7] Yoichi M 2006 Applications of piezoelectric actuator *NEC Technical Journ*al 82-86
[8] Jin X and Xie B 2015 Experimental study on surface generation in vibration-assisted micro-milling of glass *International Journal of Advanced Manufacturing Technology* **81** 507-512
[9] Abdullah A, Shahini M and Pak A 2009 An approach to design a high power piezoelectric ultrasonic transducer *Journal of Electroceramics* **22** 369-382
[10] Xu J and Huanhuan R 2017 Design and finite element simulation of an ultrasonic transducer of two piezoelectric discs Journal of Measurements in Engineering 5 266-272
[11] Kurosawa M K, Kodaira O, Tsuchitoi Y and Higuchi T 1998 Transducer for high speed and large thrust ultrasonic linear motor using two sandwich-type vibrators *IEEE Transactions on Ultrasonics, Ferroelectrics, and Frequency Control* **45** 1188-1195
[12] Kurniawan R, Ko T J, Ping L C, Kumaran S T, Kiswanto G, Guo P and Ehmann K F 2017 Development of a two-frequency, elliptical-vibration texturing device for surface texturing *Journal of Mechanical Science and Technology* **31** 3465-3473

IOP Conf. Series: Materials Science and Engineering **654** (2019) 012016 doi:10.1088/1757-899X/654/1/012016

Investigation of Structure and Tensile Properties of Betel Palm Sheath

C Choonamchai[1], D Panwised[1], P Mitsomwang[1] and R Borrisutthekul[1]

[1]School of Metallurgical Engineering, Suranaree University of Technology, 111 University avenue, Muang, Nakhon Ratchasima, 30000, Thailand

pusit@sut.ac.th

Abstract. Betel palm sheath is a kind of natural material which is increasingly used as a raw material for producing food packaging. To form packaging such as food plates, a raw palm sheath is drawn using drawing dies. Then, the excess edge of the drawn part is cut off with trimming dies. Due to the lack of understanding of the palm sheath structure and its mechanical properties, a lot of defective parts still occur during the production. In this work, the authors aim to investigate the internal structure of the palm sheath. The raw palm sheath was sectioned and subjected to a scanning electron microscope (SEM). In addition, the mechanical properties of the palm sheath were examined by the uni-axial tensile test. Based on the experimental results, the internal structure of the palm sheath was clearly revealed. The tensile properties of the palm sheath were highly anisotropic, namely the maximum load resistance and the elongation at break remarkably varied with the loading direction. Moreover, the deformation of the palm sheath appeared to consist of the elastic and the permanent deformation.

1. Introduction

Betel palm shown in Figure 1 is a kind of plants in the Arecaceae family which is widely found in South east Asia such as India, Myanmar, Thailand etc. Parts such as nut, husk and sheath of the betel palm can be used in different proposed. For example, its wet sheath is used as animal feed, while the dry palm sheath can be used as a fuel for cooking and water heating [1-3].

Figure 1. Betel palm.

According to the plastic packaging waste issue, the production of food packaging from natural material is now intensively considered. The betel palm sheath is a kind of natural materials which has

Content from this work may be used under the terms of the Creative Commons Attribution 3.0 licence. Any further distribution of this work must maintain attribution to the author(s) and the title of the work, journal citation and DOI.
Published under licence by IOP Publishing Ltd

potential to be used for producing the food packaging because it has sheet form, suitable thickness and sufficiently strength. In Thailand, Veerasa company utilises the palm sheath to produce various packaging such as food plate, banana-cake cup, soap box, etc [4]. The production starts from the preparation of raw palm sheath by washing, drying and cutting into designed size. Then, the palm sheath is drawn using stainless steel drawing dies at elevated temperature. After drawing, the edge of the drawn part is trimmed to obtain finishing products. However, a lot of defective parts still occur during the production. This seems to be caused by the lack of understand of the structure and the mechanical properties of the palm sheath.

There are researchers studying mechanical properties of natural materials-based composites. Srinivasan et al., studied the mechanical and thermal properties of banana-flax based natural fibre composite. They applied the hand layup method for preparing the composite laminates. The results showed that the difference lamination strongly affected the ultimate tensile strength, the tensile elongation, the energy absorption and fire resistance of the composite [5]. Fernandes et al., prepared the natural composite structures between the flax fiber, the agglomerated cork and the bio-resin. They found that the composite had higher load resistance when the density of the fiber in the composite increased [6]. In addition, Chethan et al., prepared the betel palm leaf reinforced resin composite and investigated its tensile properties. Their composite consisted of 70% Epoxy resin and 30% short palm sheath fiber (by volume). They revealed that the ultimate tensile strength and the modulus of the composite was 16.15 MPa and 3345.16 MPa, respectively [7].

From the literature survey, the authors found that there are not any study or report explaining the structure and the mechanical properties of the betel palm sheet. To promote the production of the palm sheath packaging from the household level to the industrial scale, the fundamental knowledge concerning the structure and the mechanical properties of the palm sheet is highly required. Therefore, in this study, the betel palm sheath was collected, sectioned and subjected to the SEM in order to investigate the internal structure of the palm sheath. Also, the palm sheath tensile specimens were prepared from different zones and directions in a palm sheath. Through the experimental results, the structure, the tensile properties and the deformation behavior of the palm sheath were discussed and summarized.

2. Material and experimental conditions

Betel palm sheath used in this study was harvested from Buriram prefecture, Thailand. For all of the investigation, only one palm sheath which had the width and the length of 300 and 400 mm was used. The moisture content of the palm sheath was 16.15%. The entire palm sheath was cut off using a keen knife to prepare the SEM and the uni-axial tensile specimens based on ASTM D828-16.

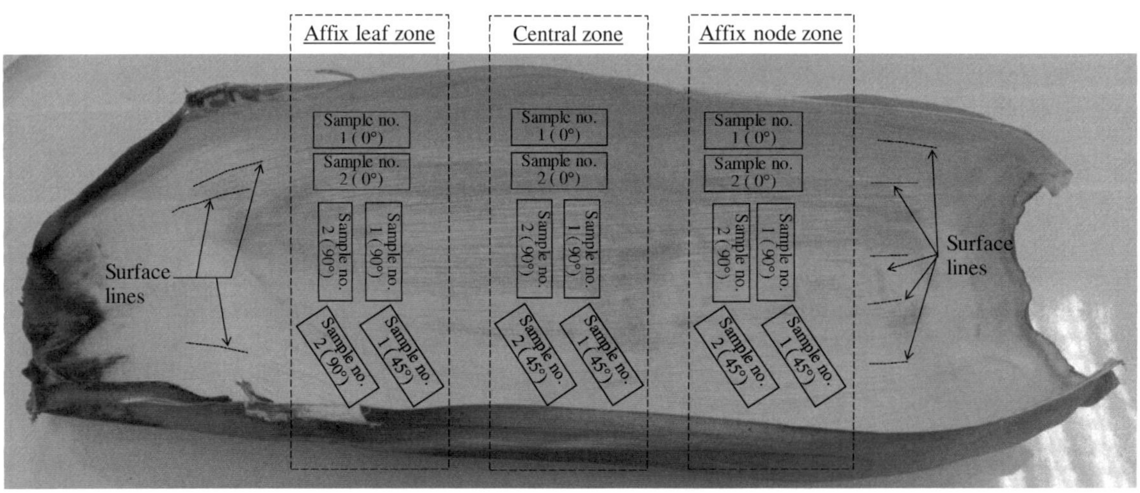

Figure 2. Layout of tensile specimens.

Before testing, the thickness of the cut samples was measured by a micrometre. At the affix leaf zone, the thickness of the palm sheath varied ranging from 1.972 to 2.526 mm (Avg. = 2.354 mm). For the central zone, the thickness was varied from 1.429 to 2.000 mm (Avg. = 1.718 mm), while at the affix node zone, the thickness was varied between 1.318 mm and 2.255 mm (Avg. = 1.818 mm). Based on the variation of the thickness, the tensile testing samples were formed into three groups for those three zones. In each zone, the tensile specimens were prepared in three directions with respect to the surface lines of the palm sheath. The first direction was parallel to the surface line (They were called "0° specimens"). The second and the third directions were diagonal and perpendicular to the surface lines (They were called "45° and 90° specimens, respectively"). Figure 2 shows the layout of the tensile specimens. For the tensile test, the speed of the moving cross head was fixed as 2 mm·min^{-1}. The load cell (Capacity: ± 5 kN) was installed to record the load resistance of the palm sheath. After testing, the fracture edges of the specimens were investigated with the SEM.

Apart from the tensile test mentioned above, additional specimens were prepared for studying the deformation behavior of the palm sheath. For this test, the specimens were cut from the affix node zone, prepared to have the dimension based on the same standard as that used for tensile test. Then, small grids were drawn on the surface of the specimens using a small-head permanent pen. After that the specimens were subjected to the uni-axial tensile load. The displacement of the cross head was varied from 0.2 to 1.6 mm. After testing, the displacement-load resistance curve and the geometry of the grids were examined.

3. Results and discussions

3.1. Structure of betel palm sheath

Figure 3 shows the cross-sectional SEM images of the palm sheath. From this image, six different components of the palm sheath, i.e. spongy mesophyll, phloem, xylem, bundle sheath, upper and lower epidermis were clearly observed. Figure 4 represents the high magnification SEM images at the upper and the lower surface of the palm sheath. These revealed that the upper epidermis was thinner than the lower epidermis. Namely, the upper one had the thickness of about 2.5 μm, while the thickness of the lower one was approximately 20 μm.

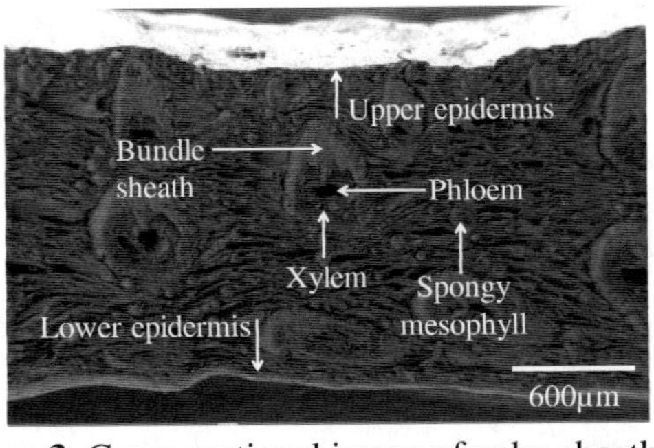

Figure 3. Cross-sectional image of palm sheath.

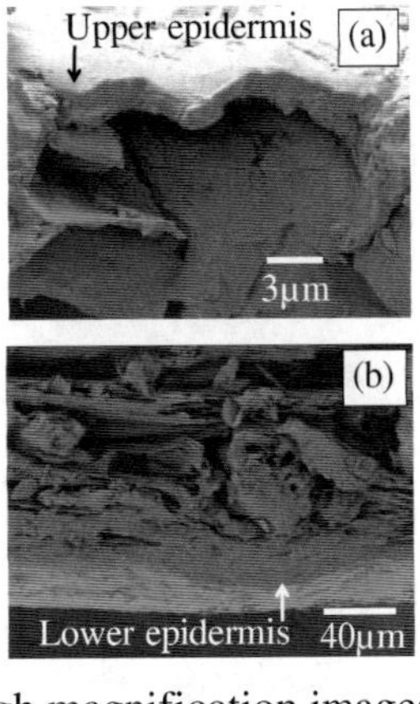

Figure 4. High magnification images (a) at upper and (b) at lower surfaces.

3.2. Tensile properties of betel palm sheath

After carrying out the tensile test, the relationship between the load resistance of the palm sheath and the extension was plotted. Figure 5 represents the load resistance curves for all the tensile specimens. The load resistance characteristics were revealed as follows: (i) the highest maximum load resistance ($f_{Max.}$) occurred when loading direction was parallel to the surface line of the palm sheath, 0° direction. Its value was ranging from 309 to 405 N, while the highest $f_{Max.}$ was varied between 62 and 106 for the

IOP Conf. Series: Materials Science and Engineering **654** (2019) 012016 doi:10.1088/1757-899X/654/1/012016

45 and 90° loading directions. (ii) There was a variation of the highest $f_{Max.}$ when varying the investigation zone of the palm sheath. However, the tendency of this variation could not be observed. (iii) The elongation at break (e_{Break}) remarkably depended on the loading direction rather than the investigation zone of the betel palm sheath. In the cases of 0 and 90° loading directions, the e_{Break} appeared to be low (between 1.0 to 1.9 mm), while the e_{Break} was higher in the case of 45° loading direction. From the above tensile testing results, it was revealed that tensile properties of the palm sheath were highly anisotropic.

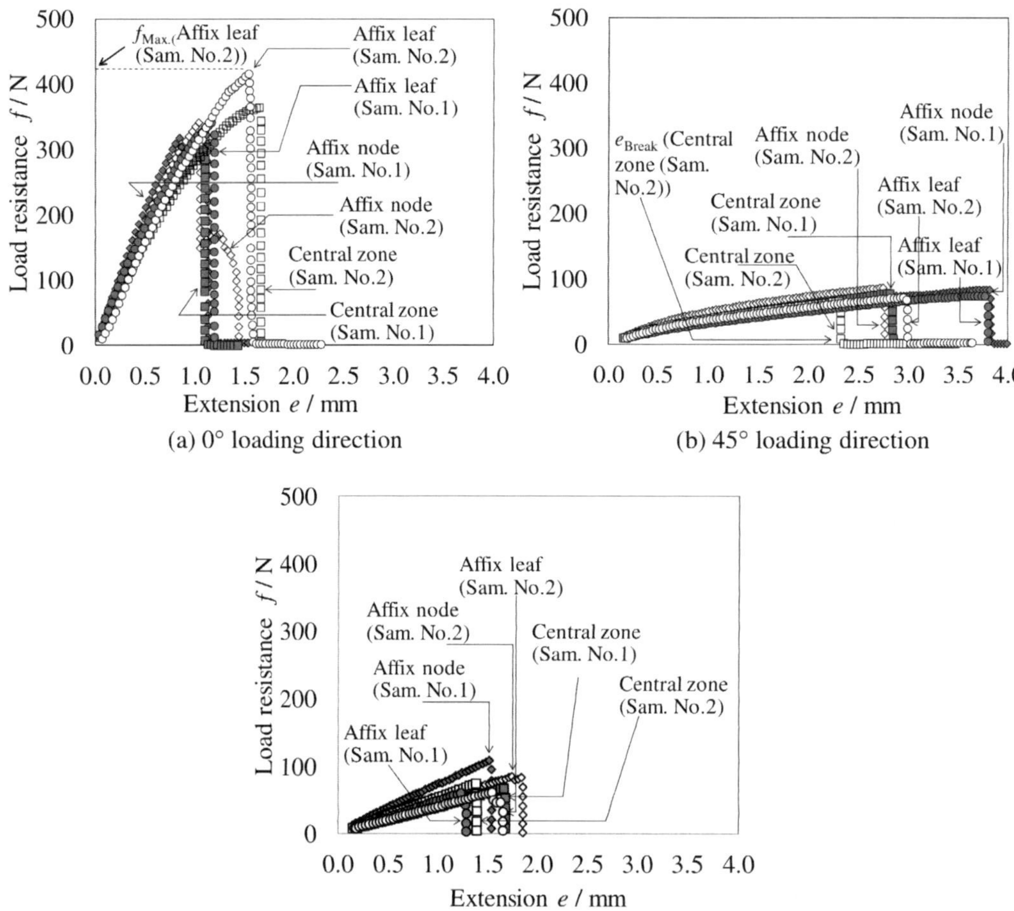

(a) 0° loading direction (b) 45° loading direction

(c) 90° loading direction

Figure 5. Relationship between load resistance and extension of palm sheath from tensile test.

3.3. Fracture edges of tensile specimens

After tensile testing, the fractured edges of the specimens were investigated with the SEM. Figure 6 shows the SEM images of the fractured edges which were loaded under 0, 45 and 90° at the affix leaf, the central and the affix node zones. This figure revealed that (i) For all zones of the palm sheath, in the case of 0°, the xylem and phloem appeared to be broken. But, they tended to be pulled out from the spongy mesophyll without the breakage in the cases of 45 and 90°. (ii) For all the investigation zones of the palm sheath, there was no crack propagation to the upper surface in the case of 0°, while in the cases of 45 and 90°, the crack propagation was observed. (iv) the fracture of the spongy mesophyll was similar for all specimens. It was indicated that the tensile load response of the spongy mesophyll was similar for all zones of the palm sheath and the loading directions. (v) Applying the tensile load in the 45 and 90°, the fracture of the specimens seemed to be initiated at the surface line. Considering this fracture initiation position and the load resistance curve in Figure 5, the surface line became a

IOP Conf. Series: Materials Science and Engineering **654** (2019) 012016 doi:10.1088/1757-899X/654/1/012016

week point for the fracture, especially when the maximum principal tensile stress occurred in 45 and 90° against the surface line. (vi) Seeing the high e_{Break} in figure 5(b), although the surface line caused the fracture under the low load resistance, it did not result in the low elongation of the palm sheath.

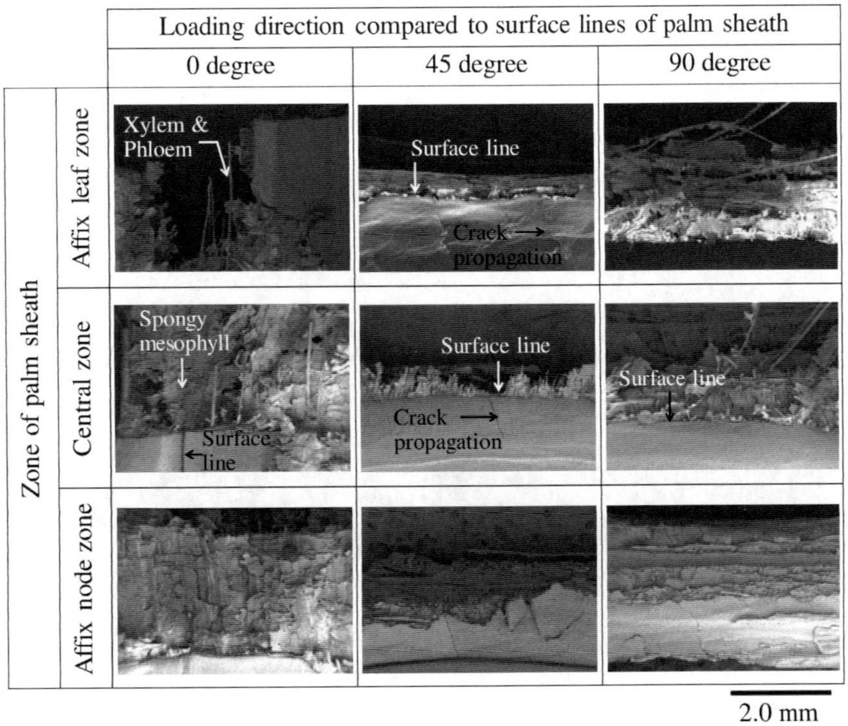

Figure 6. Fracture edge of tensile specimens.

3.4. Deformation behavior of betel palm sheath

Apart from the internal structure investigation and the tensile testing, the deformation behavior of the palm sheath was studied. This understanding is crucial, especially for analyzing the deformation of the palm sheath using numerical method such as finite element method (FEM). Typically, when assigning a material model to a simulation software, two types of the deformation modes, i.e. the elastic deformation and the plastic deformation must be assumed, separately [8].

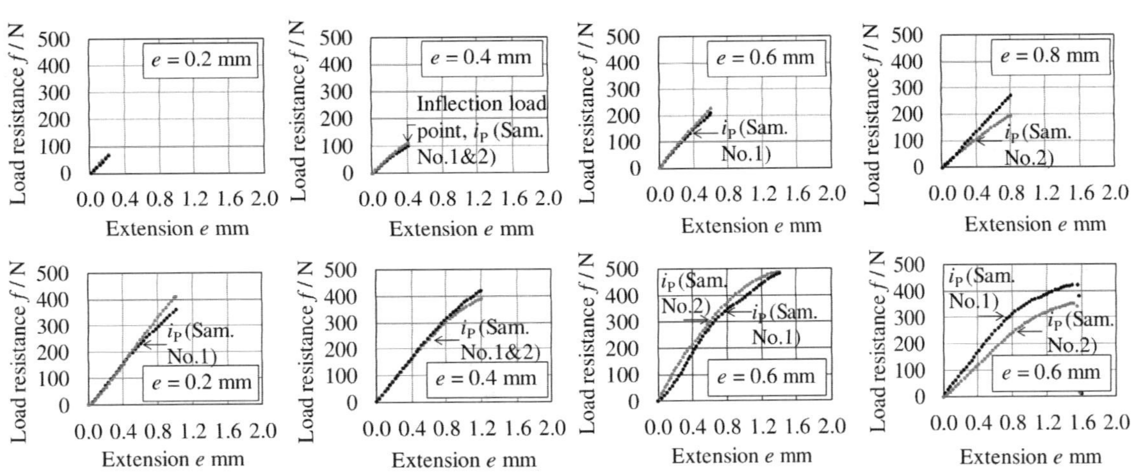

Figure 7. Relationship between load resistance and extension for extension from 0.2 to 1.6 mm.

Figure 7 shows the load resistance curves after pulling the palm sheath specimens to the specified extension. From this figure, at the small extension, $0 < e < 0.4$, the load resistance tended to be increased without changing the slope of the relationship. When the extension increased, the inflection point of the load resistance (i_P) was detected. Here, the i_P occurred at the extension between 0.4 and 0.8 mm. After pulling under the large extension, $e > 0.8$ the slope of the load resistance curve changed, noticeably. This seemed to indicate that the deformation behavior of the palm sheet was changed with increasing the extension.

Figure 8 shows the grids on the palm sheath specimens after pulling. Here, it could be seen that the grids did not deform largely in all extension cases. However, measuring the grid distance (d_G) after pulling the samples, the d_G did not change for $0.2 < e \leq 0.4$. But, for $e \geq 0.6$, the d_G tended to be extended, compared to the original d_G. This confirmed that, under the tensile loading, the betel palm sheath consisted of the elastic deformation and the permanent deformation.

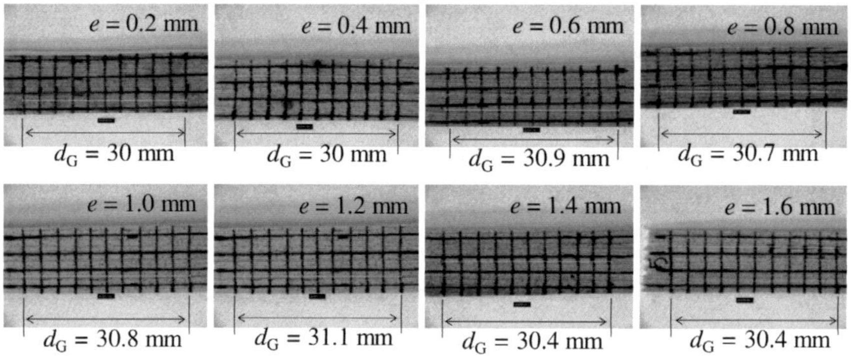

Figure 8. Palm sheath specimens after pulling at various extension

4. Conclusions

From the experimental investigation of the betel palm sheath, the following conclusions were obtained: (i) the internal structure of the palm sheath was clearly observed. It consisted of bundle sheath, phloem, xylem, spongy mesophyll, upper and lower epidermis. (ii) the tensile properties of the palm sheath were highly anisotropic. The highest maximum load resistance occurred when loading direction was perpendicular with the surface line of the palm sheath. (iii) The fracture initiation seemed to be occurred at the surface line of the palm sheath. (iv) The load resistance of the palm sheath was very low, when the maximum principal stress occurred in the 45 and 90° against the surface line of the palm sheath. (v) The palm sheath showed both the elastic deformation and the permanent deformation under the tensile loading.

5. References

[1]Shashikumar, Shrinivasa D.J., Manjunatha K. and Anantachar, M. 2016 *Int. J. of Agriculture Sciences* **8(6)** 701
[2]Bhattarai, BN. *Climate-smart Agriculture (CSA) 1*, 1
[3]Nikhil R., Raahul N., Sateesh N., Shivpriyan M., Tiffany C., Anita T. and Pramod R. 2018 *Materials Today : Proceedings* **5** 24541
[4]Veerasa palm sheath packaging. (2019) Available: http:// http://www.veerasa.com/th
[5]Srinivasan V.S. Boopathy R., Sangeetha D. and Ramnath B.B. 2014 *Materials and Design* **60** 620
[6]Fernandes F.A.O., Tavares J.P., Alves de Sousa R.J., Perira A.B. and Esteves J.L. 2017 *Procedia Manufacturing* **13** 227
[7]Chethan M.R., Gopala Krishna Dr.S.G., Chennakeshava R. and Mahesh D. 2016 *Int. J. of Engineering Research and Advanced Technology* **2(1)** 388
[8]Mitsomwang P. and Nagasawa S. 2015 J. of Materials Processing Technology **220** 46

IOP Conf. Series: Materials Science and Engineering **654** (2019) 012017 doi:10.1088/1757-899X/654/1/012017

Lithium Salt Doped Poly(Vinylidene Fluoride)/Cellulose Acetate Composite Gel Electrolyte Membrane for Lithium Ion Battery

Muhammad Rehman Asghar[a], Muhammad Tuoqeer Anwar[ab], Tahir Rasheed[c], Ahmad Naveed[d], Xiaohui Yan[a] and Junliang Zhang[a]*

[a] Institute of Fuel Cells, School of Mechanical Engineering, Shanghai Jiao Tong University, MOE Key Laboratory of Power Machinery and Engineering, Shanghai Jiao Tong University, Shanghai China.
[b] Comsat University Islamabad (Sahiwal Campus), off G.T Rd., Sahiwal, Punjab 57000, Pakistan.
[c] School of Chemistry and Chemical Engineering. State key laboratory of metal matrix composites, shanghai Jiao Tong University, Shanghai 200240.
[d] Shanghai Electrochemical Energy Devices Research Center, School of Chemistry and Chemical Engineering, Shanghai Jiao Tong University, Shanghai 200240.

E-mail: junliang.zhang@sjtu.edu.cn

Abstract. In this study, composite gel electrolyte based on Polyvinylidene fluoride (PVDF)/cellulose acetate with the addition of different proportions of lithium nitrate ($LiNO_3$) is prepared by a simple casting method. Salt doping effects on gel electrolyte membrane is investigated by scanning electron microscopy, X-ray analysis, DSC, FT-IR and TGA analysis. Salt doped gel electrolyte membrane shows a porous structure with a high uptake of 150%. Furthermore, with the addition of 20 wt% salt in the gel electrolyte membrane increase the ion conductivity from 0.21×10^{-3} S cm^{-1} to 3.0×10^{-3} S cm^{-1} at room temperature after soaking 1M $LiPF_6$ EC/DEC liquid electrolyte. As compared with PP 2325 separator, pristine PVDF and PVDF/cellulose acetate gel electrolyte membrane, the salt doped PVDF/cellulose acetate membranes exhibit wide electrochemical window and lower interfacial resistance. Composite gel electrolyte membrane with 10 wt% salt contents possess 100% capacity retention after 100 cycles at 1C rate as compared to PP 2325 separator which shows only 87% capacity retention.

1. Introduction

Lithium ion battery with high charge density, long cycling, high thermal stability and high resistible against chemical reactions with electrolyte is highly desirable [1][2]. Lithium ion battery consists of cathode, anode and separator. The separator is the key component which is responsible to enhance the safety of the battery by separating the positive and negative electrode as well as provide a smooth path for the movement of lithium ions. Commercial polyfin separator (PE, PP and PP/PE/PP) Polypropylene (PP), are most widely used separators in lithium based batteries which consist of thickness in the range of 10-25μm. However, their low thermal stability due to low melting temperature and low ion conductivity due to hydrophobic in nature still make them insufficient for advanced lithium ion batteries (LIBs) [3].

Nowadays, gel polymer electrolyte is a promising candidate as a separator for application of lithium ion batteries due to its high ion conductivity, the capability of holding electrolyte without leakage, chemically inert and electrochemical stable for both electrodes. These properties make it superior to solid

Content from this work may be used under the terms of the Creative Commons Attribution 3.0 licence. Any further distribution of this work must maintain attribution to the author(s) and the title of the work, journal citation and DOI.
Published under licence by IOP Publishing Ltd

polymer electrolytes and liquid electrolytes [4,5]. Various polymer hosts of gel polymer electrolytes have been studied such as polymethylmethacrylate (PMMA)[6], Polyimide [7], polyacrylonitrile (PAN) [8], polyetherimide[9], polyvinylidene fluoride (PVDF) and its copolymers [10,11]. Polyethylene oxide (PEO) [12]. Among them, PVDF holds a special place. PVDF has a strong electron-withdrawing functional group (–C–F) and the high dielectric constant($\varepsilon = 8.4$) which is beneficial for the larger dissolution of lithium salts and for a higher concentration of charge carriers. But due to its low crystallinity, hydrophobic in nature and low ion conduction provokes it's application in lithium ion batteries [13].

Cellulose acetate is on the promising derivatives of cellulose which can easily soluble in an organic and polar solvent. CA possess high glass transition temperature, electrically non-conductor and high affinity towards organic electrolyte due to acetyl (C=O) and hydroxyl group (-OH) make it a promising candidate for LIBs [14].

Lithium nitrate ($LiNO_3$) is a low cost lithium salt which is using as a filler and to develop the tunable structure for lithium ion battery separators. Due to its high electronegativity and charge delocalization, it enhances the ion conduction of the membrane and also has a great ability to make complex such as cellulose acetate [15], PVDF[16] etc. In this work, PVDF and cellulose acetate blended gel electrolyte membrane (GEM) is developed which dopes with lithium nitrate in the backbone of cellulose acetate. The effects of lithium nitrate on morphology, ion conductivity and cycling performance of the lithium ion battery were investigated.

2. Experimental

2.1. Materials

Cellulose acetate (CA)[CAS:9004-35-7] was purchased from Sigma Aldrich , Poly (vinilidene fluoride) (PVDF) keynar 761 was purchased from Alchemy chemical company , N, N-Dimethylformamide (DMF) was purchased from Aladdin, Lithium nitrate [$LiNo_3$] was purchased from Chemical reagent Co. Ltd, The electrolyte 1M $LiPF_6$ solution in EC/DEC solvent which was supplied by MTI corporation.

2.2. Experimental

The schematic diagram was shown in Figure 1. PVDF and Cellulose acetate polymer (ratio 4:1) was dissolved into DMF solvent at 70 °C for 1h. After a homogeneous solution, certain amounts of lithium nitrate was added into a solution and further stirred for 23 h at 70 °C. Then the solution was cast on a Teflon plate and dried at 50 °C to form composite gel electrolyte membrane. After drying, the solid electrolyte membrane was activated in a liquid electrolyte (Figure 1). The different weight percent salt (0, 10, 20,) containing electrolyte membrane was designated by M_1, M_{10}, M_{20} and pristine PVDF membrane. The thickness of the samples was set to 128-134 μm. Note that a further increase of lithium salt amount from 20 wt% makes the membrane mechanically weak. For comparison purpose, pristine PVDF, PVDF/cellulose acetate (M_1) commercial Celgard PP 2325 separator was dipped into electrolyte separately.

2.3. Preparation of cathode electrode

The Lithium iron phosphate based cathodes were made by blade coating method in which 80 % cathode material, 10 % black carbon and 10 % PVDF in NMP solvent were mixed together by stirring for 12 h. Then the slurry was cast on the plain aluminum sheet attached on a clean glass plate. Then these sheets were dried for 12 h under vacuum at 120 °C temperature. After heating, aluminum sheet cut into a disc of 14 mm diameter. The lithium disc of 16 mm diameter was selected as an anode.

2.4. Characterizations

The surface of composite gel electrolyte membranes with and without salt were examined by field emission scanning electron microscopy (FE-SEM, Hitachi S-4800, Japan), and samples were sputter-coated with gold before characterization.

The liquid electrolyte uptake was obtained by immersing the salt doped gel polymer electrolyte membranes in 1M$LiPF_6$ EC+DEC electrolyte for 2h. The surface electrolyte was gently removed by filter paper. The liquid electrolyte uptake was calculated by the following formula.

$$Electrolyte\ Uptake = \frac{W_w - W_d}{W_d}\ x\ 100 \quad (1)$$

Where W_d is the weight of the dry membrane and W_w is the weight of the membrane after immersing into the liquid electrolyte. To confirm the readings, 5 pieces of each sample were tested and the mean value was taken.

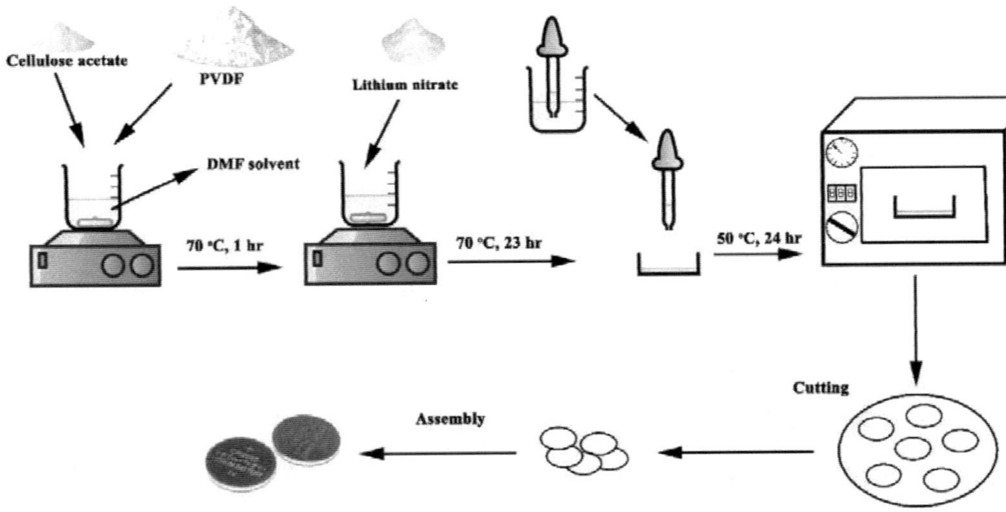

Figure 1. Preparation method of gel electrolyte membrane based on salt for lithium ion battery.

Fourier transform infrared (FT-IR) spectra of the samples were recorded using IR/Nicolet 6700 Fourier transform infrared spectrometer in the range of 600~4000 cm^{-1}. Differential scanning calorimetry (DSC) characterization was performed using Differential scanning calorimeter (DSC 8000) by Perkin Elmer under a nitrogen atmosphere from 50 to 200 °C at a heating rate of 10 °C min^{-1}. X-ray diffraction (XRD) patterns were recorded using out using a Bruker AXS D8 Advance X-ray diffractometer with Cu Kα radiation (1.540 Å) in the 2θ range between 5 and 85 ° at 5°C per min.

The ionic conductivities of S-GPEs were measured with blocking-type cell, in which the S-GPE was sandwiched between two stainless steel (SS) electrodes. Impendence data were obtained from AUTOLAB PGSTAT302N using frequency range from 1 to 10^5 Hz and potential amplitude of 5 mV. The ionic conductivity was obtained by the following equation.

$$Ion\ conductivity\ (I) = \frac{T}{RxA} x\ 100 \quad (2)$$

Where T and A represent the thickness and the area of the circular shaped membrane sample, and *R* corresponds to impedance resistance.

For activation energy calculation, ion conductivity was measured from 30 to 90°C temperature. The following formula was used to get the activation energy value.

$$Activation\ energy\ (E_a) = -slope\ x\ R \quad (3)$$

Where slope value was obtained from a graph using origin (version 8.5) and R is the universal gas constant which has a value of 8.31416KJ mol^{-1}.

IOP Conf. Series: Materials Science and Engineering **654** (2019) 012017 doi:10.1088/1757-899X/654/1/012017

Linear sweep voltammetry (LSV) was measured by sandwich the gel electrolyte membrane sample between stainless steel electrode (working electrode) and lithium metal (Reference electrode) with a potential range of 0.2–6.0 V (vs. Li/Li^+) at a scanning rate of 5 mV s^{-1}.

The salt Li/S-GPE/LFP cells were tested on Neware Charge-Discharge testing equipment at 25 °C temperature. The cells were cycled in the voltage range of 3.0–4.2 V at different charge densities (0.2 to 8C rate) and for 100 cycles at 1C rate.

3. Results and discussions

Table 1. Uptake, ion conductivity values and related resistance values for composite gel electrolyte

Sr. no.	Ion conductivity ($mS\ cm^{-1}$) at 25 °C	Electrolyte uptake (%)	Resistance (ohm)	E_a Kj mol^{-1}
Pure PVDF	0.21	55	30.36	10.03
M_o	1.17	107.33	5.50	7.87
M_{10}	1.56	136	3.99	6.42
M_{20}	3.0	150	2.24	4.693
Celgard 2325	0.57	120%	2.51	7.53

Solution casting method usually creates a flexible, high quality mechanically strong film. The morphology of the PVDF membrane along with cellulose acetate blend and lithium salt doped polymers blended membrane is shown in Figure 2a-d. The results show that PVDF and PVDF/cellulose acetate membrane possess dense top layer with irregular pores shape due to the high affinity of PVDF and DMF solvent[Figure 2(a, b)] [17]. Lithium salt disrupts the polymer chain due to high dissociation with polymers and creates a porous structure which is beneficial for high electrolyte uptake and high ion conductivity [Figure 2(c, d)].

The electrolyte uptake was measured by immersing the gel electrolyte membrane in 1M $LiPF_6$ EC: DEC solvent. Table 1 shows that the electrolyte uptake values for pristine PVDF, PVDF/cellulose acetate and Gel polymer membrane with 10 wt% and 20 wt% salt contents are 50%, 107.33%, 139% and 150% respectively. The reason behind low uptake of PVDF is due to dense and less porous structure which does not contribute to the absorption of electrolyte [18]. The salt dopes gel electrolyte membrane exhibits high electrolyte uptake as compared to pristine PVDF, PVDF/cellulose acetate and commercial PP 2325 separator because of its good computability with electrolyte and possesses porous structure which holds the electrolyte.

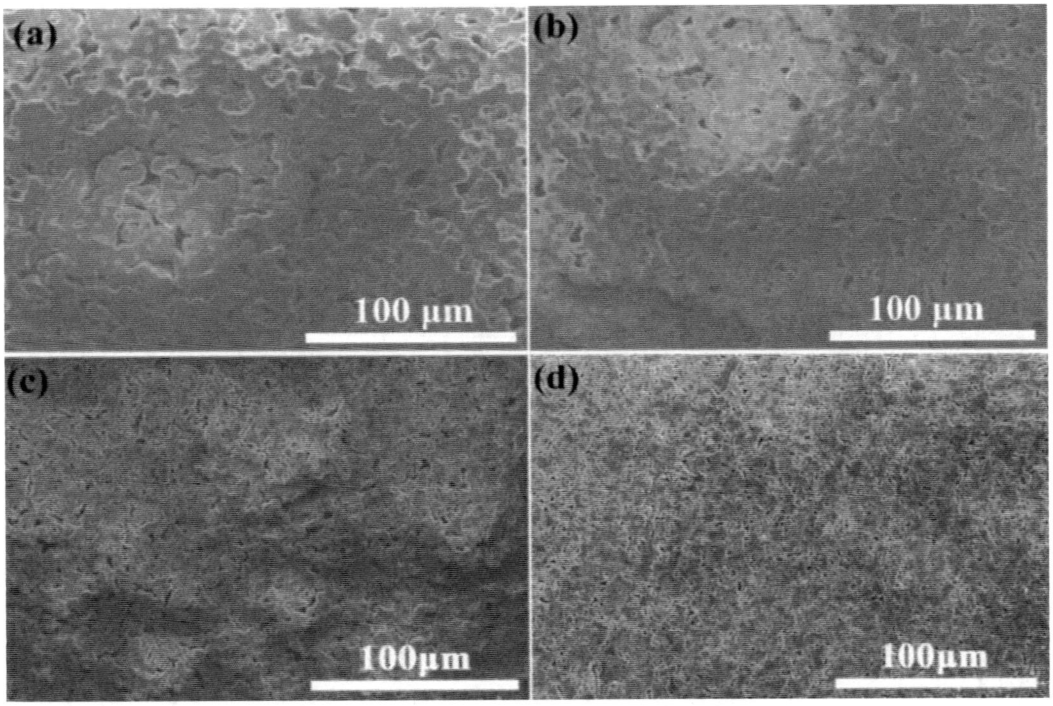

Figure 2. SEM of (a) Pristine PVDF, (b) PVDF/cellulose acetate (M_o) (c) 10 wt% salt doped gel electrolyte membrane (d) 20 wt% salt doped gel electrolyte membrane.

To check the interactions between lithium nitrate, cellulose acetate and PVDF FT-IR spectroscopy were applied. Figure 3(a) shows FT-IR bands of pristine PVDF, PVDF/cellulose acetate (M_1), 10 and 20% salt doped membrane (M_{10}, M_{20}). The bands 908, 1030, 1219, 1371, and 1736 cm^{-1} represent CH_2 rocking, C–O–C stretching, C–O stretching, CH_2 bending, and C=O symmetric stretching, respectively. The change in intensity of the peaks represents the successful introduction of the salt in the polymers complex. The band at 3420 cm^{-1} observe after addition of lithium salt which is due to moisture absorbed by lithium nitrate due to its hygroscopic nature [19]. Moreover, the stretching of the carbonyl group at 1749 cm^{-1} is due to strong intermolecular interaction between the oxygen atom of the carbonyl group and lithium ion which result salt polymer complex [20]. The FT-IR results show no extra bands appear with the addition of salt which indicates that there is no chemical interaction between salt and polymers blends.

The crystallinity of the gel electrolyte membrane is measured by XRD pattern. Figure 3(b) shows the XRD peaks of pristine PVDF, PVDF/cellulose acetate and 10 and 20% salt doped PVDF/cellulose acetate membrane. The peaks at 2θ=18.3°, 20.2° and 25.6° represent pristine PVDF membrane and peaks at 2θ= 16.3° and 22.5° are the diffraction peaks of the Cellulose acetate. After addition of lithium nitrate $LiNo_3$ the intensity of the peaks substantially decline which indicates the successful dissociation of salt with polymers blend and also an indication of increment in amorphous nature of gel electrolyte membrane by decreasing the crystallinity of the gel electrolyte membrane.

To further analysis the crystallinity change in gel polymer electrolyte membrane, DSC analysis was employed. DSC is a thermal analysis technique to measure the heat flow which passes in or out of the sample as a function of temperature while the sample is experienced controlled temperature.

Figure 3(c) shows DSC curves of the Pristine PVDF, PVDF/cellulose acetate (M_1) and salt doped membranes. The values of heat of fusion and melting temperature of polymer blends and salt doped membranes are listed in table 2. The Table 1 shows that, with the addition of cellulose acetate and salt amounts, the melting temperature shifts from 172.2 °C to 167.5 °C and heat of fusion decline from 42.57

$J.g^{-1}$ to 24.27 $J.\ g^{-1}$, which indicates the successful formation of polymers and salt complex. The results also show that the addition of lithium salt creates free volume and amorphous region in the PVDF matrix.

Table 2. DSC characterization values of composite electrolyte membranes

Sample Name	Pristine PVDF	M_1	M_{10}	M_{20}
T_m (°C)	172.2	170	169	167.5
ΔH_f ($J.g^{-1}$)	42.57	38.44	34.57	24.27
Crystallinity (%)	35	24	16	11

TGA curves of Pristine PVDF, PVDF/cellulose acetate (M_1) and salt doped composite electrolyte membrane is shown in the Figure 3d shows that the Pristine PVDF membrane decomposes at 470.5.0°C with 60% weight loss. The PVDF/cellulose acetate decomposes into two stages. In one stage the decomposition occurs at 334°C with the weight loss of 18.5% which is due to decomposition of cellulose acetate while in the second stage the decomposition occurs at 483.2 °C with 80% weight loss. After the addition of salt, decomposition occurs at 487.2 °C which is also the decomposition temperature of lithium nitrate.

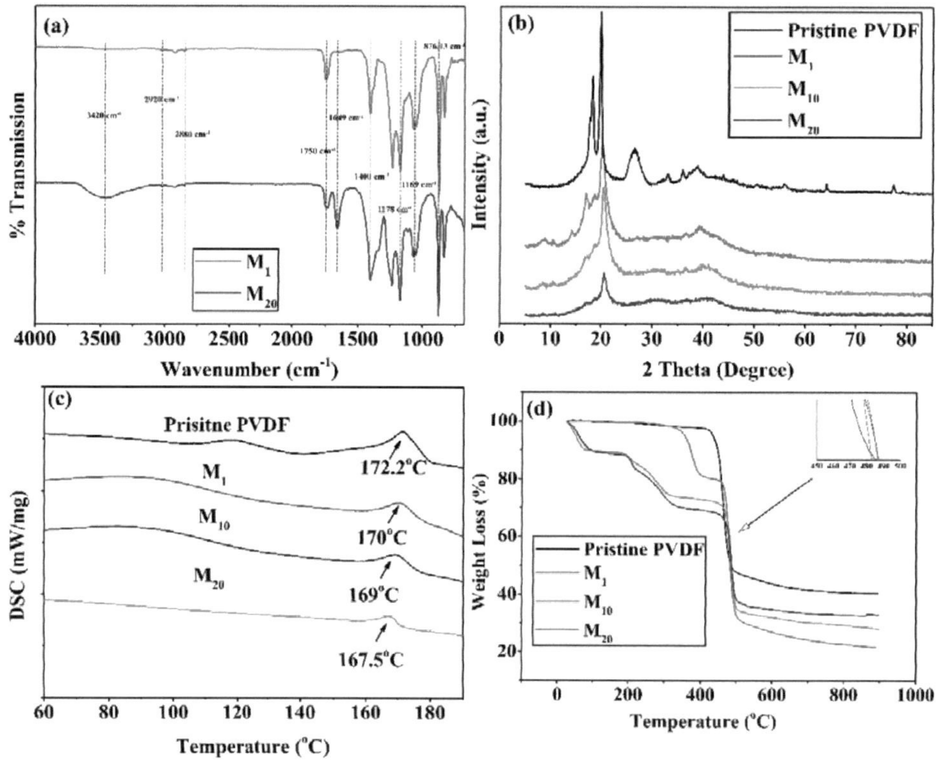

Figure 3. (a) FT-IR spectra profile of PVDF/cellulose acetate membrane and 20 wt% salt doped gel electrolyte membrane (b) XRD curves of composite gel electrolyte membrane with and without salt dope. (c) DSC curves of Pristine PVDF, PVDF/cellulose acetate, M_{10} and M_{20} salt doped gel electrolyte membrane. (d) TGA curves of pristine PVDF, PVDF/cellulose acetate (M_1) and salt doped gel electrolyte membranes.

The Results demonstrate that the salt doped composite electrolyte membrane decomposition temperature is higher than pristine PVDF and PVDF/cellulose blend.

Cellulose acetate is easily soluble in electrolyte due to high solubility in organic solvents such as ethyl methyl carbonate (EMC) and polycarbonate (PC) [21]. To test the stability of the gel electrolyte membrane containing cellulose acetate, it dips into the electrolyte solution for 2 and 24h as shown in Figure 4(a-d). The results show that there is no distortion of membrane due to the electrolyte which confirms strong hydrogen bonding exists between PVDF and cellulose acetate and gel electrolyte membrane can be used as a separator for lithium ion battery.

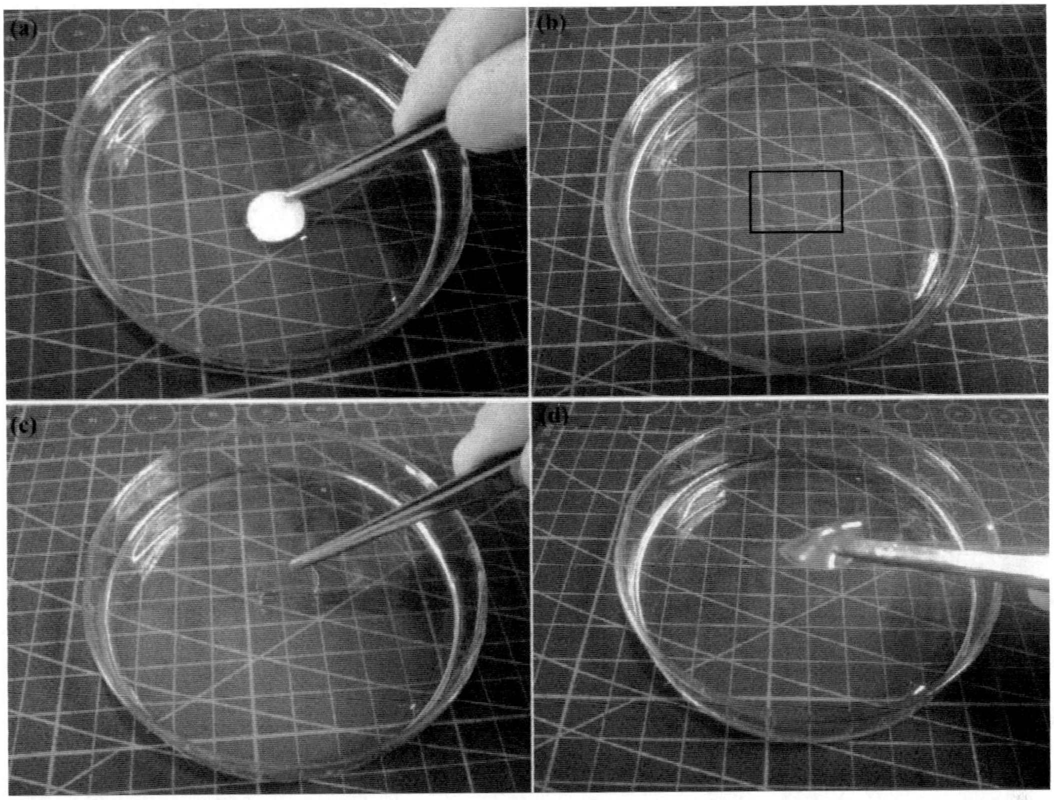

Figure 4. (a) Dip into electrolyte 1M $LiPF_6$ (EC+DEC) (b) view after 2h (c) view after 24h (d) membrane flexibility.

The results also show that the salt does not influence the hydrogen bonding between cellulose acetate and PVDF. Figure 4(d) shows that even after 24h remain in the electrolyte, the gel electrolyte membrane is flexible enough to bear the pressure inside a coin cell without structure damage.

Generally, the ionic conductivity in conventional battery separator is achieved due to the continuous pathways developed by the absorbed liquid electrolyte within the interconnected pores of the membrane. The porous structure is one of the important factors in determining the ionic properties of an electrolyte in a separator. Also of importance, the ionic conductivity of a porous membrane undoubtedly depends on the conductivity of the electrolyte embedded in the pores of the membrane. Figure 5(a) shows AC impedance of different GEM. The ion conductivity values of pristine PVDF, pristine PVDF/cellulose and membrane with 10 wt% and 20 wt% lithium nitrate are 0.21, 1.17, 1.56, and 3.00 mS cm^{-1} which is higher than the previous research [22]. The enhanced ion conductivity of salt doped gel electrolyte membrane is due to high electrolyte uptake and adhesion towards electrode which reduces the interfacial resistance. Salt doped gel electrolyte membrane still contains high electrolyte uptake as compared to commercial PP separator (0.57 mS cm^{-1}). Figure 5(b) shows the Arrhenius plots of the ion conductivity for gel electrolyte membrane with and without salt doping. With increasing temperature ion conductivity increase which shows that the curves exactly obeys the polymer electrolyte Arrhenius behavior [23][24]. Table 1 shows that the activation energy (E_a) values for Pristine PVDF, PVDF/cellulose acetate (M_o), commercial PP 2325 separator and salt doped composite electrolyte membrane with 10 and 20% salt are 10.03 KJ mol^{-1}, 7.83 KJ mol^{-1}, 6.42 KJ mol^{-1} and 4.693 KJ mol^{-1} respectively which calculated from the formula is given in the reference [25]. The activation energy values of Pristine PVDF and PVDF/cellulose acetate (M_o) is high because they expand during heating and create free volume for the migration of ions through their polymer structure[26]. By combining the salt with gel electrolyte membrane, the expansion of the polymer cease which results in low activation energy.

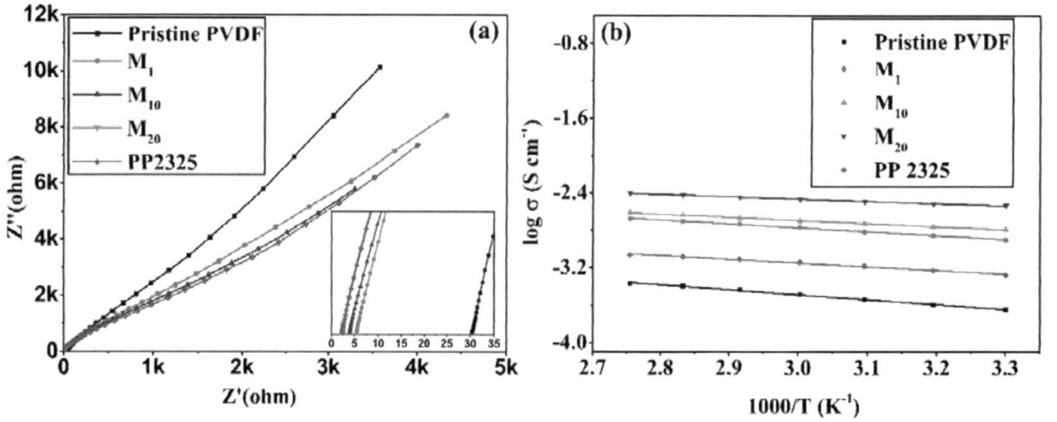

Figure 5. (a) AC impedance spectra of GEM at 25 °C (b) Arrhenius plots of ion conductivities of GEM.

Linear sweep voltammetry (LSV) use to investigate electrical window of separators. Generally, the electrical window for lithium ion battery is limited to 2 to 4.3V. Figure 6a shows that the Pristine PVDF and PVDF-CA (M_1) decomposition temperature is 4.5V which is also a decomposition voltage of liquid electrolyte [27]. After addition of salt electrical window increase up to 4.65V, the results show that salt doped gel electrolyte membrane possess good electrochemical stability and can be used as a replacement of commercially available separator in lithium ion battery.

The EIS spectrum is measured by sandwiched the gel polymer electrolyte membrane between two lithium metal. Usually, impedance curves consist of two sections as shown in Figure 6(b). Section one is the straight line in the low frequency range which represents charge transfer resistance (R_b) and section two is a semicircular shape which represents the diffusion resistance (R_i) [28]. Figure 6(b) shows that the pristine PVDF and PVDF/cellulose acetate(M_o) contain high interfacial resistance because of the reaction between the F atom of PVDF and electrode during charging and discharging process [29]. After the addition of the lithium salt, the interfacial resistance substantially decline. It is concluded that the salt doped gel electrolyte membranes show soft contact with lithium anode to maintain small interfacial impedance and adoption of volume change in the electrode.

The electrochemical performance of the salt doped gel electrolyte membrane with Pristine PVDF, PVDF/cellulose acetate (M_1), salt doped composite gel electrolyte membranes M_{10}, M_{20} was investigated with Li anode and lithium iron phosphate (LFP) material as a cathode. Lithium iron phosphate is a very promising candidate with a stable capacity for next generation lithium ion battery applications. Figure 6(c) shows a charge-discharge profile in the potential window of 3.0 to 4.2V at different C rate (0.2C to 8C) using different composite gel electrolyte membrane. In general, the discharge values decrease with increasing charge density due to lithium iron phosphate inside lithium diffusion by extraction/intercalation reaction behavior [30]. PP, PVDF and PVDF/cellulose acetate (M_o) exhibit less capacity retention at high charge density while salt doped gel electrolyte membrane possesses high capacity even at high charge density because of their superior ion conductivity, high electrolyte uptake. Figure 6(d) shows a discharge capacity of 10 wt% salt doped gel electrolyte membrane and PP 2325 separator at 1C rate for 100 cycles. The commercial separator shows 88% capacity retention after 100 cycles with 118 mA h g^{-1} initial and

100mA h g^{-1} final discharge capacity values while 10 wt% salt doped gel electrolyte deliver 100% capacity retention after 100 cycles with approx. 125 mA h g^{-1} initial and final discharge capacity values respectively. The superior discharge capacity retention is due to high electrolyte uptake and ion conductivity of salt doped gel electrolyte membrane.

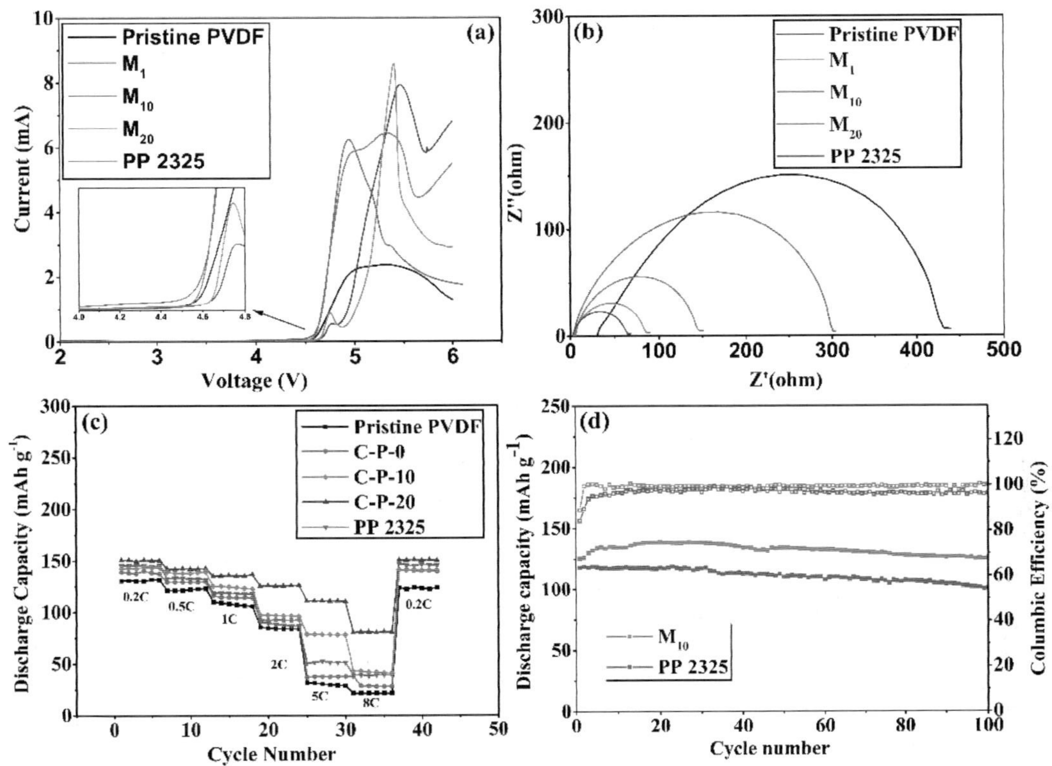

Figure 6. (a) The electrochemical window of Pristine PVDF, PVDF-CA (M_1) and 10 and 20 wt.% salt doped gel electrolyte membranes (M_{10}, M_{20}) (b) Interfacial Impedance curves of Pristine PVDF, PVDF/cellulose acetate (M_o), 10 wt% and 20 wt% salt doped gel electrolyte membranes. (c) Discharge capacity at different charge density (d) Cycling performance of 10 wt% salt doped gel electrolyte membrane and PP 2325 separator at 1C rate.

4. Conclusions

Lithium salt doped PVDF/cellulose acetate electrolyte membrane has been developed through the casting method. Due to the high affinity with cellulose acetate and PVDF polymer, lithium salt increased the physical and electrochemical properties of the blended membrane. Due to high affinity towards electrolyte, salt doped electrolyte membrane shows high ion conductivity. Lithium salt doped gel electrolyte possesses good electrochemical window of 4.65V. With 10 wt% salt doping with gel electrolyte membrane provide 100% capacity retention at 1C rate with 100% columbic efficiency. In short, salt doped PVDF/cellulose acetate composite gel electrolyte membrane is a promising candidate as a separator for rechargeable lithium ion batteries.

Acknowledgments

This work was supported by PSA Peugeot Citroën-SJTU Project (No. 13H100000584), National Natural Science Foundation of China (No. 21406220), and SJTU-UM Project (No. 15X120010002).

5. References

[1]Gong S-D, Huang Y, Cao H-J, Lin Y-H, Li Y, Tang S-H, Wang M-S and Li X 2016 A green and environment-friendly gel polymer electrolyte with higher performances based on the natural matrix of lignin *J. Power Sources* **307** 624–33

[2]Imholt L, Dörr T S, Zhang P, Ibing L, Cekic-Laskovic I, Winter M and Brunklaus G 2019 Grafted polyrotaxanes as highly conductive electrolytes for lithium metal batteries *J. Power Sources* **409** 148–58

[3]Moon J, Jeong J Y, Kim J Il, Kim S and Park J H 2019 An ultrathin inorganic-organic hybrid layer on commercial polymer separators for advanced lithium-ion batteries *J. Power Sources* **416** 89–94

[4]Xu K 2014 Electrolytes and interphases in Li-ion batteries and beyond *Chem. Rev.* **114** 11503–618

[5]Long L, Wang S, Xiao M and Meng Y 2016 Polymer electrolytes for lithium polymer batteries *J. Mater. Chem. A* **4** 10038–9

[6]Krejza O, Velická J, Sedlaříková M and Vondrák J 2008 The presence of nanostructured Al2O3 in PMMA-based gel electrolytes *J. Power Sources* **178** 774–8

[7]Wang H, Wang T, Yang S and Fan L 2013 Preparation of thermal stable porous polyimide membranes by phase inversion process for lithium-ion battery *Polymer (Guildf).* **54** 6339–48

[8]Evans T, Lee J H, Bhat V and Lee S H 2015 Electrospun polyacrylonitrile microfiber separators for ionic liquid electrolytes in Li-ion batteries *J. Power Sources* **292** 1–6

[9]Zhai Y, Xiao K, Yu J and Ding B 2015 Fabrication of hierarchical structured SiO 2 /polyetherimide-polyurethane nanofibrous separators with high performance for lithium ion batteries *Electrochim. Acta* **154** 219–26

[10]Saito Y, Kataoka H, Capiglia C and Yamamoto H 2000 Ionic Conduction Properties of PVDF - HFP Type Gel Polymer Electrolytes with Lithium Imide Salts *J. Phys. Chem. B* **104** 2189–92

[11]Asghar M R, Zhang Y, Wu A, Yan X, Shen S, Ke C and Zhang J 2018 Preparation of microporous Cellulose/Poly(vinylidene fluoride-hexafluoropropylene) membrane for lithium ion batteries by phase inversion method *J. Power Sources* **379** 197–205

[12]Li H, Ma X T, Shi J L, Yao Z K, Zhu B K and Zhu L P 2011 Preparation and properties of poly(ethylene oxide) gel filled polypropylene separators and their corresponding gel polymer electrolytes for Li-ion batteries *Electrochim. Acta* **56** 2641–7

[13]Li H, Chen Y-M, Ma X-T, Shi J-L, Zhu B-K and Zhu L-P 2011 Gel polymer electrolytes based on active PVDF separator for lithium ion battery. I: Preparation and property of PVDF/poly(dimethylsiloxane) blending membrane *J. Memb. Sci.* **379** 397–402

[14]Xiao S, Wang F, Yang Y, Chang Z and Wu Y 2014 An environmentally friendly and economic membrane based on cellulose as a gel polymer electrolyte for lithium ion batteries *RSC Adv.* **4** 76–81

[15]Monisha S, Mathavan T, Selvasekarapandian S, Benial A M F and Latha M P 2017 Preparation and characterization of cellulose acetate and lithium nitrate for advanced electrochemical devices *Ionics (Kiel).* **23** 2697–706

[16]Janakiraman S, Surendran A, Ghosh S, Anandhan S and Venimadhav A 2019 A new strategy of PVDF based Li-salt polymer electrolyte through electrospinning for lithium battery application *Mater. Res. Express* **6** 035303

[17]Susan L Y, Ismail S, Ooi B S and Mustapa H 2017 Surface morphology of pvdf membrane and its fouling phenomenon by crude oil emulsion *J. Water Process Eng.* **15** 55–61

[18]Shi X, Ma N, Wu Y, Lu Y, Xiao Q, Li Z and Lei G 2018 Fabrication and electrochemical properties of LATP/PVDF composite electrolytes for rechargeable lithium-ion battery *Solid State Ionics* **325** 112–9

[19]Eswaraiah V, Sankaranarayanan V and Ramaprabhu S 2011 Inorganic nanotubes reinforced polyvinylidene fluoride composites as low-cost electromagnetic interference shielding materials *Nanoscale Res. Lett.* **6** 137

[20]Selvakumar M and Krishna Bhat D 2008 LiClO4 doped cellulose acetate as biodegradable polymer electrolyte for supercapacitors *J. Appl. Polym. Sci.* **110** 594–602

[21]Chen W, Zhang L, Liu C, Feng X, Zhang J, Guan L, Mi L and Cui S 2018 Electrospun Flexible Cellulose Acetate-Based Separators for Sodium-Ion Batteries with Ultralong Cycle Stability and Excellent Wettability : The Role of Interface Chemical Groups *ACS Appl. Mater. Interfaces* **10** 23883–90

[22]Cui J, Liu J, He C, Li J and Wu X 2017 Composite of polyvinylidene fluoride–cellulose acetate with Al(OH)3 as a separator for high-performance lithium ion battery *J. Memb. Sci.* **541** 661–7

[23]Xiao S Y, Yang Y Q, Li M X, Wang F X, Chang Z, Wu Y P and Liu X 2014 A composite membrane based on a biocompatible cellulose as a host of gel polymer electrolyte for lithium ion batteries *J. Power Sources* **270** 53–8

[24]Wu C, Lu M and Chuang H 2005 PVdF-HFP/P123 hybrid with mesopores: a new matrix for high-conducting, low-leakage porous polymer electrolyte *Polymer (Guildf).* **46** 5929–38

[25]Asghar M R, Zhang Y, Wu A, Yan X, Shen S, Ke C and Zhang J 2018 Preparation of microporous Cellulose/Poly(vinylidene fluoride-hexafluoropropylene) membrane for lithium ion batteries by phase inversion method *J. Power Sources* **379** 197–205

[26]Cheng Q, Cui Z, Li J, Qin S, Yan F and Li J 2014 Preparation and performance of polymer electrolyte based on poly(vinylidene fluoride)/polysulfone blend membrane via thermally induced phase separation process for lithium ion battery *J. Power Sources* **266** 401–13

[27]Kang W, Ma X, Zhao H, Ju J, Zhao Y, Yan J and Cheng B 2016 Electrospun cellulose acetate/poly(vinylidene fluoride) nanofibrous membrane for polymer lithium-ion batteries *J. Solid State Electrochem.* **20** 2791–803

[28]Zhang Z, Kong L-L, Liu S, Li G-R and Gao X-P 2017 A High-Efficiency Sulfur/Carbon Composite Based on 3D Graphene Nanosheet@Carbon Nanotube Matrix as Cathode for Lithium-Sulfur Battery *Adv. Energy Mater.* **7** 1602543

[29]Scrosati B, Croce F and Panero S 2001 Progress in lithium polymer battery R&D *J. Power Sources* **100** 93–100

[30]Lung-Hao Hu B, Wu F-Y, Lin C-T, Khlobystov A N and Li L-J 2013 Graphene-modified LiFePO4 cathode for lithium ion battery beyond theoretical capacity *Nat. Commun.* **4** 1687

Institute of Physics
Dirac House, Temple Back
Bristol BS1 6BE UK

CURRAN ASSOCIATES INC.
proceedings.com

ISSN: 1757-8981
ISBN 978-1-5108-9849-3
9 781510 898493